Sensors for Robots II

Sensors for Robots II

Editors

Xin Zhao
Mingzhu Sun
Qili Zhao

MDPI

Basel • Beijing • Wuhan • Barcelona • Belgrade • Novi Sad • Cluj • Manchester

Editors
Xin Zhao
College of Artificial Intelligence
Nankai University
Tianjin
China

Mingzhu Sun
College of Artificial Intelligence
Nankai University
Tianjin
China

Qili Zhao
College of Artificial Intelligence
Nankai University
Tianjin
China

Editorial Office
MDPI AG
Grosspeteranlage 5
4052 Basel, Switzerland

This is a reprint of articles from the Special Issue published online in the open access journal *Sensors* (ISSN 1424-8220) (available at: https://www.mdpi.com/journal/sensors/special_issues/sensor_robot_II).

For citation purposes, cite each article independently as indicated on the article page online and as indicated below:

Lastname, A.A.; Lastname, B.B. Article Title. *Journal Name* **Year**, *Volume Number*, Page Range.

ISBN 978-3-7258-1675-0 (Hbk)
ISBN 978-3-7258-1676-7 (PDF)
doi.org/10.3390/books978-3-7258-1676-7

Cover image courtesy of Xin Zhao, Mingzhu Sun and Qili Zhao

Contents

About the Editors

Xin Zhao

Xin Zhao (Professor) received his B.S. degree in Control Theory and Control Engineering from Nankai University, Tianjin, China, in 1991, his M.S. degree in Control Theory and Control Engineering from the Shenyang Institute of Automation, CAS, Shenyang, China, in 1994, and his Ph.D. in Control Theory and Control Engineering from Nankai University in 1997. In 1997, he joined Nankai University, where he is currently a Professor and the Dean of the College of Artificial Intelligence. His research interests include micromanipulation, microsystems, and mathematical biology.

Mingzhu Sun

Mingzhu Sun (Member, IEEE) received the B.S. degree in computer science and technology, the M.S. degree in computer application, and the Ph.D. degree in control theory and control engineering from Nankai University, Tianjin, China, in 2003, 2006, and 2009, respectively. She joined the faculty at Nankai University in 2009, she is currently a Professor and the Dean of Intelligent Science Department. Her research interests are in micromanipulator, image processing, and computer vision.

Qili Zhao

Qili Zhao (Associate Professor) received his Ph.D. in Control Theory and Control Engineering from Nankai University, Tianjin, China, in 2014. He was a Post-Doctoral Researcher with the Robotics and Mechatronics Research Laboratory, Department of Mechanical and Aerospace Engineering, Monash University, Melbourne, VIC, Australia, from 2014 to 2015 and a Post-Doctoral Fellow with the Advanced Micro and Nanosystems Laboratory, Department of Mechanical and Industrial Engineering, Toronto University, Toronto, ON, Canada, from 2015 to 2018. He is currently an Associate Professor at the College of Artificial Intelligence, Nankai University. His current research interests include robotic patch clamps, robotic cell manipulation, and robotic cell measurement.

sensors

MDPI

Editorial

Sensors for Robots

Xin Zhao [1,2,*], Mingzhu Sun [1,2] and Qili Zhao [1,2]

1 National Key Laboratory of Intelligent Tracking and Forecasting for Infectious Diseases, The Engineering Research Center of Trusted Behavior Intelligence, Ministry of Education, The Tianjin Key Laboratory of Intelligent Robotics (tjKLIR), Institute of Robotics and Automatic Information System (IRAIS), Nankai University, Tianjin 300350, China; sunmz@nankai.edu.cn (M.S.); zhaoqili@nankai.edu.cn (Q.Z.)

2 The Institute of Intelligence Technology and Robotic Systems, Shenzhen Research Institute of Nankai University, Shenzhen 518083, China

* Correspondence: zhaoxin@nankai.edu.cn

1. Introduction

Currently, robots are playing significant roles in industry [1], life sciences [2], education [3], medicine [4], social services [5], and the military [6]. Robots will undoubtedly play increasingly important roles in the future life of human beings [7]. Sensors are key components of robots, enabling them to perceive additional environmental information, just like the scanners of the famous R2D2 Robot in Star Wars movies [8]. They are the basis of the self-adaptive abilities and automation control abilities of robots. Therefore, novel sensing techniques and advanced sensor applications for robots have received increasing attention worldwide [9].

Novel sensor technologies for robotics have been developed to detect new environmental parameters or improve the measurement accuracy of the present parameters [10]. To achieve this, novel design [11], fabrication [12], and calibration methods [13] have been invented to improve the detection sensitivity of the sensors. Besides the aforementioned sensor innovations, advancements in the integration [14], diffusion [15], and processing methods [16] of the detected signals have been made to improve the SNR and acquire useful information from gross signals. Based on these new technologies, novel force [17], vision [18], tactile [11], and auditory [19] sensors have been reported in recent years.

With the rapid development of swarm robots and multi-agent systems, multi-sensors [20], reconfigurable sensors [21], and cyber–physical sensing systems [22] have been developed to detect the parameters of each robot or agent. Because each body of the above system is usually far from the other, wireless sensor networks [23], multi-sensing intelligent systems [20], and sensor systems for human–robot interactions [24] have been developed to facilitate the communication and processing of the signals of each robotic body.

With the widespread application of robots, sensors for novel robotic applications have made significant progress. For example, vision sensors and vision algorithms for motive object tracking and robot navigation have been widely applied in mobile robots [25], unmanned aerial vehicles [26], and underwater vehicles [27]. In addition to the applications in traditional research areas, robot sensing technologies have also been widely applied in interdisciplinary areas, including life science [2], material sciences [28], physical sciences [29], and micro-nano manipulations [30].

2. Overview of Contribution

In this context, this topical collection includes ten papers focusing on the latest advancements in the field of sensors for robots and robotic sensing applications. Each of the ten original contributions accepted for publication in this Special Issue has undergone a rigorous review process involving a minimum of two expert reviewers across at

Citation: Zhao, X.; Sun, M.; Zhao, Q. Sensors for Robots. *Sensors* **2024**, *24*, 1854. https://doi.org/10.3390/s24061854

Received: 27 February 2024
Accepted: 5 March 2024
Published: 14 March 2024

least two rounds of revision. The studies featured in the current topical collection are summarized as follows.

In Contribution 1, the authors proposed a multi-path scattering model for virtual vegetation. On this basis, a passive microwave imaging algorithm was introduced to reconstruct vegetation images based on the scattering data from the multi-path model received by the UAV (Unmanned Aerial Vehicle) robot. The combination of the electromagnetic scattering model and image processing method contributes to understanding the image results and the multi-path scattering mechanisms of vegetation, which provide a reference for the research and development of microwave imaging systems of UAV robot networking using satellite transmitting signals.

In Contribution 2, the authors proposed a novel reinforcement learning-based hierarchical control framework to enable a snake robot with an onboard camera to realize autonomous self-localization and path following. A series of simulations and hardware experimental results validate that the proposed method is capable of achieving precise and fast convergence for the path-following tasks for a snake robot with autonomous visual perception.

Contribution 3 by Yang et al. discussed the coverage of practical models with varying intensities. This study analyzed the properties of the effective coverage of multi-node teams consisting of a given number of nodes. Numerical analyses and simulations for 3-node and n-node teams ($n \geq 4$) were conducted separately. For the 3-node cases, the relations between the side lengths of equilateral triangle formation and the effective coverage of the team equipped with two different types of models were inspected. For the n-node cases ($n \geq 4$), the effective coverage of a team in three formations, namely regular polygon, regular star, and equilateral triangular tessellation (for $n = 6$), were investigated. This study provides useful insights and guides for the development of practical devices for coverage applications.

In Contribution 4 by Imtiaz et al., the target was to have the agent learn to perform prehensile and non-prehensile robotic manipulations to improve the efficiency and throughput of the pick-and-place task. To achieve this target, this study specified the problem as a Markov decision process (MDP) and deployed a deep reinforcement learning (RL) temporal difference model-free algorithm known as the deep Q-network (DQN). Experiments carried out under a range of complex scenario test cases show promising results compared to a baseline deep learning approach and a ResNet architecture-based approach.

In Contribution 5, Shi et al. presented a design and a control algorithm for a robot used for grinding the surfaces of large, curved workpieces with unknown parameters, such as wind turbine blades. First, the structure and motion mode of the grinding robot were determined. Considering that fuzzy PID has the advantages of variable parameters and strong adaptability, this study proposed a force/position hybrid control strategy based on fuzzy PID. The experimental results of robotic grinding verified the feasibility and effectiveness of the proposed control strategy compared with the artificial results.

Contribution 6 by Li et al. proposed a robotic intracellular pressure measurement method using a traditional micropipette electrode system setup, in which intracellular pressure, a key physical parameter of the intracellular environment, has been found to regulate multiple cell physiological activities and impact cell micromanipulation results. This measurement method does not necessitate specialized and expensive equipment and causes limited damage to cell viability. The experimental intracellular pressure measurement value for porcine oocytes, which is consistent with the value reported in the relevant literature, verifies the effectiveness of the proposed measurement method. Moreover, a 90% survival rate of operated oocytes was obtained after measurement, indicating limited damage to cell viability.

In Contribution 7, Isabel et al. presented a novel active device for shoulder rotation based on force control. This device is designed for shoulder rehabilitation and can perform three different exercises. This device uses a force-control architecture and an adaptive-speed PI controller. The study conducted in 12 shoulder rehabilitation sessions showed a significant improvement in shoulder rotation. The patients reported high levels of device acceptance. This device could be a valuable tool for shoulder rehabilitation and improve the efficiency of physiotherapy services.

In Contribution 8, Shu et al. proposed a method for the accurate 3D posture sensing of soft actuators. It utilizes miniaturized sponge-resistive materials and LSTM neural networks. This method considers the hysteresis and non-linear sensing signals. It was demonstrated on a 3-DOF pneumatic bellow-shaped actuator with nine sensors. The sensors exhibited a low error rate in predicting the position of the actuator tip.

In Contribution 9, Rostkowska et al. focused on appearance-based localization using catadioptric cameras. The authors proposed an efficient neural network architecture for recognizing indoor scenes from distorted images. The system was optimized for real-time inference on edge devices. Transfer learning and fine-tuning of a limited number of catadioptric images yielded the best results. The system was tested on small-scale datasets and compared with state-of-the-art systems. The proposed system performed well in terms of accuracy and inference time, providing a cost- and energy-efficient solution for appearance-based localization.

In contribution 10, Weidenbach et al. discussed the challenges of estimating the 6D poses of transparent objects for robotic manipulation. Robotic manipulation requires accurate object pose knowledge. Estimating the poses of transparent objects is challenging because of unreliable depth data and the different effects of clutter. The authors proposed a transparency-aware multi-segmentation approach and transparency-aware bounding boxes. The proposed approach improves pose estimation for transparent objects without modifying the pose estimator. The study demonstrated a 4.3% increase in the ADD-S AUC value using transparency-aware segmentation. The transparency-aware segmentation can be applied to any dataset that provides a 3D model of an object and its ground-truth pose.

3. Conclusions

In summary, this topical collection includes ten novel and interesting works in the area of sensors for robots and robotic sensing applications. We would like to express our deepest gratitude to the Managing Team of *Sensors* for their continuous support throughout the preparation of this collection. We greatly appreciate all the contributing authors and the anonymous expert reviewers for their invaluable efforts in selecting submissions of the utmost quality.

Author Contributions: Conceptualization and supervision X.Z., writing—review and editing, M.S., writing—original draft preparation, Q.Z. All authors have read and agreed to the published version of the manuscript.

Funding: This research was jointly supported by the National Natural Science Foundation of China (62027812, 62273186) and the Beijing Advanced Innovation Center for Intelligent Robots and Systems under Grant No. 2019IRS05.

Conflicts of Interest: The authors declare no conflicts of interest.

List of Contributions:

1. Chen, Z.; Qiao, X.; Wu, P.; Zhang, T.; Hong, T.; Fang, L. Unmanned Aerial Vehicle (UAV) Robot Microwave Imaging Based on Multi-Path Scattering Model. *Sensors* **2022**, *22*, 8736.
2. Liu, L.; Guo, X.; Fang, Y. A Reinforcement Learning-Based Strategy of Path Following for Snake Robots with an Onboard Camera. *Sensors* **2022**, *22*, 9867.
3. Yang, Y.-R.; Kang, Q.; She, R. The Effective Coverage of Homogeneous Teams with Radial Attenuation Models. *Sensors* **2023**, *23*, 350.
4. Imtiaz, M.B.; Qiao, Y.; Lee, B. Prehensile and Non-Prehensile Robotic Pick-and-Place of Objects in Clutter Using Deep Reinforcement Learning. *Sensors* **2023**, *23*, 1513.
5. Shi, X.; Li, M.; Dong, Y.; Feng, S. Research on Surface Tracking and Constant Force Control of a Grinding Robot. *Sensors* **2023**, *23*, 4702.
6. Li, M.; Qiu, J.; Li, R.; Liu, Y.; Du, Y.; Liu, Y.; Sun, M.; Zhao, X.; Zhao, Q. Robotic Intracellular Pressure Measurement Using Micropipette Electrode. *Sensors* **2023**, *23*, 4973.

7. Alguacil-Diego, I.M.; Cuesta-Gómez, A.; Pont, D.; Carrillo, J.; Espinosa, P.; Sánchez-Urán, M.A.; Ferre, M. A Novel Active Device for Shoulder Rotation Based on Force Control. *Sensors* **2023**, *23*, 6158.
8. Shu, J.; Wang, J.; Cheng, K.C.-C.; Yeung, L.-F.; Li, Z.; Tong, R.K.-y. An End-to-End Dynamic Posture Perception Method for Soft Actuators Based on Distributed Thin Flexible Porous Piezoresistive Sensors. *Sensors* **2023**, *23*, 6189.
9. Rostkowska, M.; Skrzypczyński, P. Optimizing Appearance-Based Localization with Catadioptric Cameras: Small-Footprint Models for Real-Time Inference on Edge Devices. *Sensors* **2023**, *23*, 6485.
10. Weidenbach, M.; Laue, T.; Frese, U. Transparency-Aware Segmentation of Glass Objects to Train RGB-Based Pose Estimators. *Sensors* **2024**, *24*, 432.

References

1. Dzedzickis, A.; Subačiūtė-Žemaitienė, J.; Šutinys, E.; Samukaitė-Bubnienė, U.; Bučinskas, V. Advanced applications of industrial robotics: New trends and possibilities. *Appl. Sci.* **2021**, *12*, 135. [CrossRef]
2. Liu, H.; Stoll, N.; Junginger, S.; Thurow, K. Mobile robot for life science automation. *Int. J. Adv. Robot. Syst.* **2013**, *10*, 288. [CrossRef]
3. Schina, D.; Esteve-González, V.; Usart, M. An overview of teacher training programs in educational robotics: Characteristics, best practices and recommendations. *Educ. Inf. Technol.* **2021**, *26*, 2831–2852. [CrossRef]
4. Dupont, P.; Nelson, B.; Goldfarb, M.; Hannaford, B.; Menciassi, A.; Yang, G. A decade retrospective of medical robotics research from 2010 to 2020. *Sci. Robot.* **2021**, *6*, eabi8017. [CrossRef] [PubMed]
5. Sheridan, T. A review of recent research in social robotics. *Curr. Opin. Psychol.* **2020**, *36*, 7–12. [CrossRef] [PubMed]
6. Gans, N.; Rogers, J. Cooperative multirobot systems for military applications. *Curr. Robot. Rep.* **2021**, *2*, 105–111. [CrossRef]
7. Pagliarini, L.; Lund, H. The future of Robotics Technology. *J. Robotics Netw. Artif. Life* **2017**, *3*, 270–273. [CrossRef]
8. Karreman, D.; Ludden, G.; Evers, V. Beyond R2D2: Designing multimodal interaction behavior for robot-specific morphology. *ACM Trans. Hum.-Robot Interact. (THRI)* **2019**, *8*, 1–32. [CrossRef]
9. Regtien, P. Sensors for applications in robotics. *Sens. Actuators* **1986**, *10*, 195–218. [CrossRef]
10. Zhao, Y.; Xing, H.; Guo, S.; Wang, Y.; Cui, J.; Ma, Y.; Liu, Y.; Liu, X.; Feng, J.; Li, Y. A novel noncontact detection method of surgeon's operation for a master-slave endovascular surgery robot. *Med. Biol. Eng. Comput.* **2020**, *58*, 871–885. [CrossRef]
11. Lambeta, M.; Chou, P.; Tian, S.; Yang, B.; Maloon, B.; Most, V.; Calandra, R. Digit: A novel design for a low-cost compact high-resolution tactile sensor with application to in-hand manipulation. *IEEE Robot. Autom. Lett.* **2020**, *5*, 3838–3845. [CrossRef]
12. Raril, C.; Manjunatha, J. Fabrication of novel polymer-modified graphene-based electrochemical sensor for the determination of mercury and lead ions in water and biological samples. *J. Anal. Sci. Technol.* **2020**, *11*, 3. [CrossRef]
13. Li, Z.; Li, S.; Luo, X. An overview of calibration technology of industrial robots. *IEEE/CAA J. Autom. Sin.* **2021**, *8*, 23–36. [CrossRef]
14. Cui, Y.; Liu, F.; Mu, J. Integrating sensing and communications for ubiquitous IoT: Applications, trends, and challenges. *IEEE Netw.* **2021**, *35*, 158–167. [CrossRef]
15. Zhu, W.; Wang, J.; Sun, W.; Zhou, S.; He, M. Preparation of gradient hydrogel for pressure sensing by combining freezing and directional diffusion processes. *Chem. Eng. J.* **2023**, *451*, 138335. [CrossRef]
16. Krishnamurthi, R.; Kumar, A.; Gopinathan, D.; Nayyar, A.; Qureshi, B. An overview of IoT sensor data processing, fusion, and analysis techniques. *Sensors* **2020**, *20*, 6076. [CrossRef]
17. Cheng, M.; Zhu, G.; Zhang, F.; Tang, W.; Yang, J.; Zhu, L. A review of flexible force sensors for human health monitoring. *J. Adv. Res.* **2020**, *26*, 53–68. [CrossRef]
18. Halwani, M.; Ayyad, A.; AbuAssi, L.; Abdulrahman, Y.; Almaskari, F.; Hassanin, H.; Zweiri, Y. A Novel Vision-Based Multi-Functional Sensor for Normality and Position Measurements in Precise Robotic Manufacturing. *Precis. Eng.* **2024**, *88*, 367–381. [CrossRef]
19. Netzer, O.; Heimler, B.; Shur, A.; Behor, T.; Amedi, A. Backward spatial perception can be augmented through a novel visual-to-auditory sensory substitution algorithm. *Sci. Rep.* **2021**, *11*, 11944. [CrossRef]
20. Ali, A.; Abouelghar, M.; Belal, A.; Saleh, N.; Yones, M.; Selim, A.; Savin, I. Crop yield prediction using multi sensors remote sensing. *Egypt. J. Remote Sens. Space Sci.* **2022**, *25*, 711–716.
21. Alamzadeh, I.; Alexandropoulos, G.; Shlezinger, N.; Imani, M. A reconfigurable intelligent surface with integrated sensing capability. *Sci. Rep.* **2021**, *11*, 20737. [CrossRef] [PubMed]
22. Andronie, M.; Lăzăroiu, G.; Ștefănescu, R.; Uță, C.; Dijmărescu, I. Sustainable, smart, and sensing technologies for cyber-physical manufacturing systems: A systematic literature review. *Sustainability* **2021**, *13*, 5495. [CrossRef]
23. Landaluce, H.; Arjona, L.; Perallos, A.; Falcone, F.; Angulo, I.; Muralter, F. A review of IoT sensing applications and challenges using RFID and wireless sensor networks. *Sensors* **2020**, *20*, 2495. [CrossRef] [PubMed]
24. Lin, K.; Li, Y.; Sun, J.; Zhou, D.; Zhang, Q. Multi-sensor fusion for body sensor network in medical human–robot interaction scenario. *Inf. Fusion* **2020**, *57*, 15–26. [CrossRef]
25. Sanchez-Ibanez, J.; Perez-del-Pulgar, C.; García-Cerezo, A. Path planning for autonomous mobile robots: A review. *Sensors* **2021**, *21*, 7898. [CrossRef]

26. Petráček, P.; Krátký, V.; Petrlík, M.; Báča, T.; Kratochvíl, R.; Saska, M. Large-scale exploration of cave environments by unmanned aerial vehicles. *IEEE Robot. Autom. Lett.* **2021**, *6*, 7596–7603. [CrossRef]
27. Sánchez, P.; Papaelias, M.; Márquez, F. Autonomous underwater vehicles: Instrumentation and measurements. *IEEE Instrum. Meas. Mag.* **2020**, *23*, 105–114. [CrossRef]
28. Xue, L.; Yamazaki, H.; Ren, R.; Wanunu, M.; Ivanov, A.; Edel, J. Solid-state nanopore sensors. *Nat. Rev. Mater.* **2020**, *5*, 931–951. [CrossRef]
29. Li, S.; Zhang, Y.; Wang, Y.; Yin, Z.; Wang, H.; Zhang, Y. Physical sensors for skin-inspired electronics. *InfoMat* **2020**, *2*, 184–211. [CrossRef]
30. Qin, J.; Jiang, S.; Wang, Z.; Cheng, X.; Li, B.; Shi, Y.; Zhu, W. Metasurface micro/nano-optical sensors: Principles and applications. *ACS Nano* **2022**, *16*, 11598–11618. [CrossRef]

sensors

Article

Unmanned Aerial Vehicle (UAV) Robot Microwave Imaging Based on Multi-Path Scattering Model

Zhihua Chen [1,2], Xinya Qiao [3], Pei Wu [3], Tiancai Zhang [1], Tao Hong [3,4,*] and Linquan Fang [4]

1 Southwest Technology and Engineering Research Institute, Chongqing 400039, China
2 Beijing Institute of Technology, Beijing 100081, China
3 School of Electronics and Information Engineering, Beihang University, Beijing 100191, China
4 Yunnan Innovation Institute·BUAA, Kunming 650233, China
* Correspondence: hongtao@buaa.edu.cn

Abstract: Unmanned Aerial Vehicle (UAV) robot microwave imaging systems have attracted comprehensive attention. Compared with visible light and infrared imaging systems, microwave imaging is not susceptible to weather. Active microwave imaging systems have been realized in UAV robots. However, the scattering signals of geographical objects from satellite transmitting systems received by UAV robots to process imaging is studied rarely, which reduces the need of load weight for the UAV robot. In this paper, a multi-path scattering model of vegetation on the earth surface is proposed, and then the microwave imaging algorithm is introduced to reconstruct the images from the UAV robot receiving the scattering data based on the multi-path model. In image processing, it is assumed that the orbit altitude of a transmitter loaded on the satellite remains unchanged, and the receiver loaded UAV robot obtains the reflective information from ground vegetation with different zenith angles. The imaging results show that the angle change has an impact on the imaging resolution. The combination of electromagnetic scattering model and image processing method contributes to understanding the image results and the multi-path scattering mechanisms of vegetation, which provide a reference for the research and development of microwave imaging systems of UAV robot networking using satellite transmitting signals.

Keywords: UAV robot microwave imaging; multi-path scattering model; UAV networking using satellite transmitting signals

Citation: Chen, Z.; Qiao, X.; Wu, P.; Zhang, T.; Hong, T.; Fang, L. Unmanned Aerial Vehicle (UAV) Robot Microwave Imaging Based on Multi-Path Scattering Model. *Sensors* **2022**, 22, 8736. https://doi.org/10.3390/s22228736

Academic Editors: Xin Zhao, Mingzhu Sun and Qili Zhao

Received: 30 October 2022
Accepted: 10 November 2022
Published: 11 November 2022

1. Introduction

Unmanned Aerial Vehicles (UAVs) are also known as flying robots [1]. The vision system of a UAV robot has a computer-controlled imaging device that allows the robot to see and adjust its flight path accordingly, which helps the UAV robot perceive, fly, avoid obstacles, and to be successfully used for search and rescue [2]. However, the vision system of a UAV robot based on visible light and infrared are sometimes greatly affected by haze, cloud and rain. Microwaves can observe day and night, and are not affected by weather and some buildings. If the UAV robot is equipped with an active microwave imaging system, the total load will be increased. In this study, we will mainly consider UAV robot microwave imaging using the satellite transmitting signals to the target and the ground, and then scattering to the UAV sensors. If the UAV robot just receives the microwave signals to process imaging, which is not equipped with a microwave transmitting source, it will reduce the need of additional load weight for the UAV robot.

A satellite communication system can be used in UAV networking [3]. In addition, the signals transmitted by satellite communication systems receiving by UAV can not only be used in wireless transmissions, but also in processed imaging. Recently, the active radar imaging method of UAV has been researched comprehensively [4]. However, the passive imaging, not equipped with a microwave transmitting source, using multi-path scattering of ground received by UAV robot based on the satellite communication systems is rarely studied.

Furthermore, theoretical studies have shown that the use of a second receiver in addition to a conventional monostatic system, i.e., a transmitter–receiver, forming a multistatic system, could improve the retrieval performance of vegetation and soil biogeophysical variables [5,6]. The passive imaging using multi-path scattering from the ground of a UAV robot can be considered as a multistatic system. In addition, the passive imaging using multi-path scattering of ground received by UAV robot based on the satellite communication systems possesses advantages such as collecting additional information of the target, more flexibility and higher imaging resolution compared to UAV robot active radar imaging because of its separated transmitter and receiver on different platforms [7–11].

To assess the performance of bistatic SAR acquisitions for crops, L. Guerriero [12] presented a theoretical study of the microwave remote sensing of vegetated surfaces. Although the sensitivity study of the observation angle was explored preliminarily based on bistatic measurements, no conclusion about how different observation angles affect the imaging results was drawn. In order to study the bistatic scattering mechanism from the ground, S. Bellez [13] developed a volume-electric-field-integral equation based on a method of moments to simulate bistatic radar scattering from the different sets of cylinders, and then undertook an experiment to verify it. Lastly, a near-field reconstruction algorithm was applied to generate images based on both theoretical and experimental scattered fields. However, since the model is from numerical solution of volume integral equation, the research object is limited to a simple reduced-scale model composed of dielectric parallelepipeds, and thus the image results don't reflect the relatively complex environment and spatial structure of vegetation appropriately. In this paper, to study the passive imaging using multi-path scattering from ground of UAV robot based on the satellite communication systems, a multi-path scattering scattering model is developed to simulate a scattering field from the virtual three-dimensional structure of vegetation, and then scattering fields of vegetation at different frequency points and positions are processed to obtain the passive multi-path scattering images.

This paper is organized as follows. A multi-path scattering model for vegetation is developed in Section 2. Section 3 introduces the passive image processing algorithm using multi-path scattering from the ground of a UAV robot. In Section 4, simulation results are presented and reasonable analysis is given. Conclusions are drawn in Section 5.

2. Model Description

2.1. Multi-Path Scattering Model

In this study, the multi-path scattering model for virtual vegetation is established based on Distorted Born Approximation (DBA) [14]. The virtual vegetation is composed of dielectric cylinders and disks which stand for the trunk, branches and leaves, respectively. The ground surface, which is described by the standard deviation of the height, correlation length and dielectric constant, is divided into many small units equally. The total scattering fields are added coherently from all the scatterers from vegetation and ground surface.

It is assumed that there are N scatterers in the vegetation area and M ground blocks. The total scattering fields include five components, in which four components are about vegetation scatterers with transmission paths inherited from Born approximation and one component is related to ground surface, as shown in Figure 1. Direct scattering from the scatterers in the vegetation is denoted as $\overline{E}_n^{ds}(k_i, x)$. The single bounce of the scattering with a path of transmitter-scatterer-ground-receiver is denoted as $\overline{E}_n^{sg}(k_i, x)$, and the scattering with a path of transmitter-ground-scatterer-receiver is denoted as $\overline{E}_n^{gs}(k_i, x)$. Furthermore, the double bounce of the scattering with a path of transmitter-ground-scatterer-ground-receiver is denoted as $\overline{E}_n^{gsg}(k_i, x)$. Lastly, direct scattering from the ground is denoted as $\overline{E}_m^{dg}(k_i, x)$. Therefore, the total scattering field is the coherent summation of each component from all the azimuth observation positions x and wave numbers k_i, which can be expressed as

$$\overline{E}(k_i, x) = \sum_{n=1}^{N} \left(\overline{E}_n^{ds}(k_i, x) + \overline{E}_n^{sg}(k_i, x) + \overline{E}_n^{gs}(k_i, x) + \overline{E}_n^{gsg}(k_i, x) \right) + \sum_{m=1}^{M} \overline{E}_m^{dg}(k_i, x) \tag{1}$$

where n denotes the scatterer in the vegetation area, and m denotes the ground block. Each component can then be expressed as

$$\overline{E}_n^{ds}(k_i,x)=\frac{e^{ikr}}{4\pi r}e^{i\varphi_n}A_t(\theta_i,\phi_i)A_r(\theta_r,\phi_r)S_n^{ds}\overline{E}_0^i \tag{2}$$

$$\overline{E}_n^{sg}(k_i,x)=\frac{e^{ikr}}{4\pi r}e^{i\varphi_n}A_t(\theta_i,\phi_i)A_r(\theta_r,\phi_r)S_n^{sg}\overline{E}_0^i \tag{3}$$

$$\overline{E}_n^{gs}(k_i,x)=\frac{e^{ikr}}{4\pi r}e^{i\varphi_n}A_t(\theta_i,\phi_i)A_r(\theta_r,\phi_r)S_n^{gs}\overline{E}_0^i \tag{4}$$

$$\overline{E}_n^{gsg}(k_i,x)=\frac{e^{ikr}}{4\pi r}e^{i\varphi_n}A_t(\theta_i,\phi_i)A_r(\theta_r,\phi_r)S_n^{gsg}\overline{E}_0^i \tag{5}$$

$$\overline{E}_n^{dg}(k_i,x)=\frac{e^{ikr}}{4\pi r}e^{i\varphi_n}A_t(\theta_i,\phi_i)A_r(\theta_r,\phi_r)S_n^{dg}\overline{E}_0^i \tag{6}$$

where $A_t(\theta_i,\phi_i)$ and $A_r(\theta_r,\phi_r)$ denote the antenna patterns of the transmitter and receiver, respectively. (θ_i,ϕ_i) and (θ_r,ϕ_r) are the transmitting and receiving angles. φ_n is phase compensation term of the n_{th} scatterer, which is the phase shift from the local to the global coordinate system. Here $\varphi_n=(\overline{k}_i-\overline{k}_s)\cdot\overline{r}_n$, where $\overline{r}_n$ denotes position vector of the n_{th} scatterer in global coordinate system. $\overline{k}_i$ and $\overline{k}_s$ are the wave vectors of incidence and scattering wave, respectively. In addition, $\varphi_m=(\overline{k}_i-\overline{k}_s)\cdot\overline{r}_m$, where $\overline{r}_m$ denotes the position vector of the m_{th} ground block. The expression of S_n^{ds}, S_n^{sg}, S_n^{gs}, S_n^{gsg} and S_n^{dg} can be written as

$$S_n^{ds}=\overline{\overline{T}}{}_n^i\overline{\overline{F}}{}_n^s(\theta_i,\phi_i;\theta_r,\phi_r,k_i)\overline{\overline{T}}{}_n^{ir} \tag{7}$$

$$S_n^{sg}=e^{i\varphi_{n1}}\overline{\overline{T}}{}_n^i\overline{\overline{F}}{}_n^s(\theta_i,\phi_i;\pi-\theta_r,\phi_r,k_i)\overline{\overline{T}}{}_n^{rr}\overline{\overline{R}}\,\overline{\overline{T}}{}^{tr} \tag{8}$$

$$S_n^{gs}=e^{i\varphi_{n2}}\overline{\overline{T}}{}^t\overline{\overline{R}}\,\overline{\overline{T}}{}_n^r\overline{\overline{F}}{}_n^s(\pi-\theta_i,\phi_i;\theta_r,\phi_r,k_i)\overline{\overline{T}}{}_n^{ir} \tag{9}$$

$$S_n^{gsg}=e^{i\varphi_{n3}}\overline{\overline{T}}{}^t\overline{\overline{R}}\,\overline{\overline{T}}{}_n^r\overline{\overline{F}}{}_n^s(\pi-\theta_i,\phi_i;\pi-\theta_r,\phi_r,k_i)\overline{\overline{T}}{}_n^{rr}\overline{\overline{R}}\,\overline{\overline{T}}{}^{tr} \tag{10}$$

$$S_m^{dg}=e^{i\varphi_{n4}}\overline{\overline{T}}{}_m^i\overline{\overline{F}}{}_m^g(\theta_i,\phi_i;\theta_r,\phi_r,k_i)\overline{\overline{T}}{}_m^{ir} \tag{11}$$

where φ_1, φ_2, φ_3 and φ_4 are the time delay terms related to the first transmission paths as shown in Figure 1, and they can be obtained by $k_0[(|\overline{r}_t-\overline{r}_n|+|\overline{r}_r-\overline{r}'_n|)-(|\overline{r}_t-\overline{r}_n|+|\overline{r}_r-\overline{r}_n|)]$, $k_0[(|\overline{r}_t-\overline{r}'_n|+|\overline{r}_r-\overline{r}_n|)-(|\overline{r}_t-\overline{r}_n|+|\overline{r}_r-\overline{r}_n|)]$, $k_0[(|\overline{r}_t-\overline{r}'_n|+|\overline{r}_r-\overline{r}'_n|)-(|\overline{r}_t-\overline{r}_n|+|\overline{r}_r-\overline{r}_n|)]$ and $k_0[(|\overline{r}_t-\overline{g}_n|+|\overline{r}_r-\overline{g}_n|)-(|\overline{r}_t-\overline{r}_n|+|\overline{r}_r-\overline{r}_n|)]$, respectively.$\overline{r}_t$ and $\overline{r}_r$ denote the coordinate vectors of the transmitter and receiver, and $\overline{r}_n$ and $\overline{r}'_n$ denote the coordinate vectors of the scatterer and the mirror of the n_{th} scatterer. $\overline{g}$ is the coordinate vector of the m_{th} ground block. $\overline{\overline{T}}{}_n^m$ (m = i, r and t) denotes transmission matrix in incident direction, representing paths of top canopies-scatterer, ground-scatterer and top canopies-ground, respectively. $\overline{\overline{T}}{}_n^{mr}$ (m = i, r and t) denotes transmission matrix in scattering direction, representing paths of scatterer-top canopies, scatterer-ground and ground-top canopies, respectively. To obtain these transmission matrices, the vegetation area is divided into many small cubes and Foldy's approximation [15] is used in the calculation process. $\overline{\overline{R}}$ is the specular reflection of ground surface. $\overline{\overline{F}}{}_n^s$ is the scattering matrix of the n_{th} scatterer. $\overline{\overline{F}}{}_m^g$ is the scattering matrix of the m_{th} ground block, and the amplitude is calculated by integral equation model [16], while the phase is set to be random. For both $\overline{\overline{F}}{}_n^s$ and $\overline{\overline{F}}{}_m^g$, (θ_i,ϕ_i) and (θ_r,ϕ_r) are the direction angles of transmitting and receiving antenna patterns. The generalized Rayleigh-Gans (GRG) approximation is used to calculate the scattering matrix from leaves. The method of infinite cylinder approximation and the generalized GRG approximation are used to calculate the scattering from trunk and branches [17]. The reflection matrix of ground surface is calculated by Fresnel reflection equation of plane.

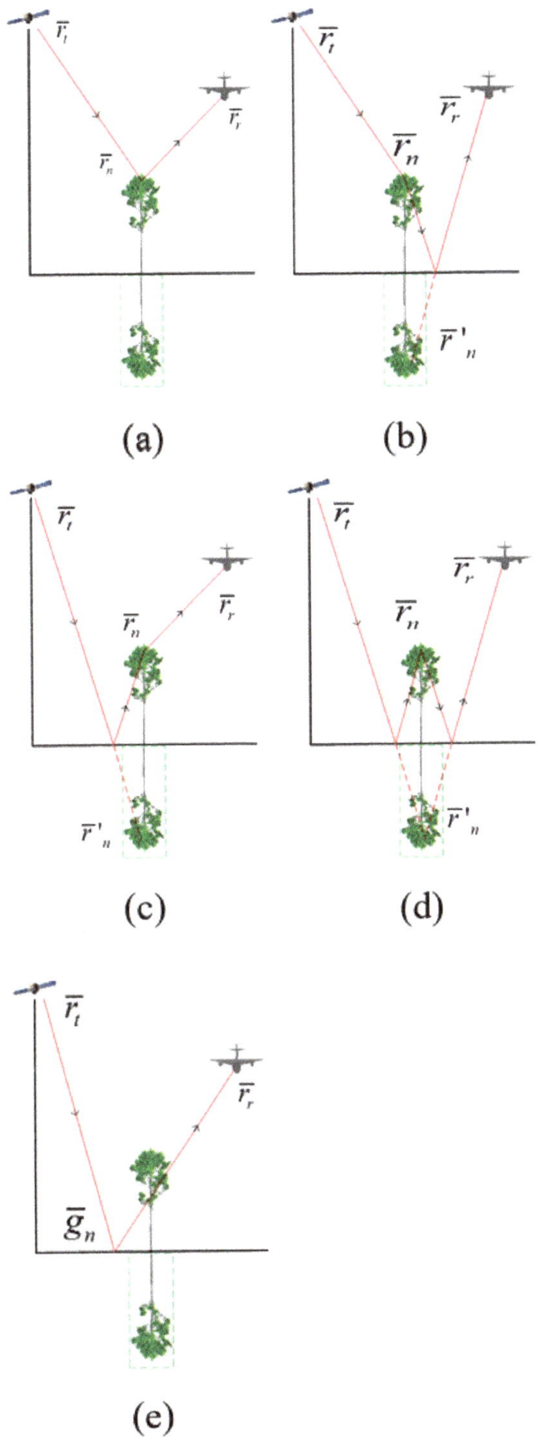

Figure 1. The five scattering components: (**a**) Direct scattering from scatterer; (**b**) Single bounce from scatterer to ground; (**c**) Single bounce from ground to scatter; (**d**) Double bounce from ground, scatter to ground; (**e**) Direct scattering from ground.

2.2. *UAV Robot Imaging Configuration Based on the Satellite Communication Systems*

A UAV robot imaging system based on the satellite communication is one with the transmitter and receiver loaded on satellite and UAV platforms, respectively, as shown in Figure 2. In this simulation, the flight tracks of the transmitter and receiver are set to be parallel, and both transmitter and receiver operate in strip mode. Furthermore, it is assumed that both transmitter and receiver work in "work-stop-work" mode based on the fact that the motion velocity of this UAV and satellite is very small compared with the signal transmitting velocity. Therefore, the satellite can transmit and the UAV robot can receive the signals at a series of discrete positions. In this study, the orbit altitude of the transmitter remains unchanged, while the heights of the receivers loaded on UAV networking are considered to obtain scattering signals at different zenith angles, which contribute to analyze the multi-path scattering mechanisms of vegetation based on UAV networking using a satellite communication system.

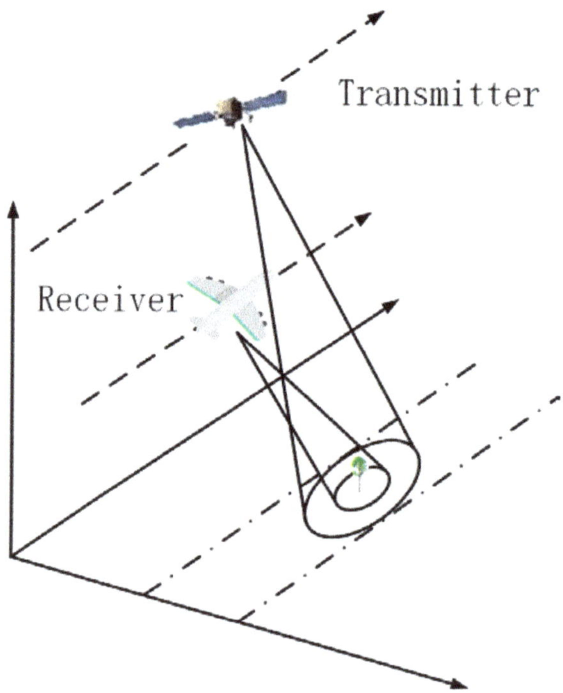

Figure 2. UAV robot observation configuration.

The working frequency is chosen at C band, and proper bandwidth is selected according to the system, as listed in Table 1. The altitude of the transmitter is 800 km, while the height of receiver is 8 km. The incident angle of the transmitting signal is 45°, and it can be demonstrated that the horizontal distance between the transmitter and the target vegtation is approximately 800 km. For the receiver, six typical zenith angles: 26°, 36°, 46°, 56°, 66° and 76° are chosen to study how the zenith angle affects the multi-path scattering imaging results. The horizontal distance between the receiver and the target vegetatioin can be calculated according to the zenith angle and flight height. Other parameters in this simulation are listed in Table 1.

Table 1. Parameters of the observation geometry.

Parameters	Value
Start frequency	6 GHz
Frequency bandwidth	300 MHz
Frequency sampling numbers	50
Flight height of the transmitter	800 km
Flight height of the receiver	8 km
Space sampling interval of transmitter	500 m
Space sampling interval of receiver	5 m
Space sampling numbers	201
Polarization	HH

3. Imaging Method

In this study, the passive imaging of the UAV robot using the satellite communication systems is developed from the field imaging algorithm [18]. The basic idea of this algorithm is to obtain the target equation from the measurement field through the inverse Fourier transform. The target equation is a function of the target position. To realize the target imaging, the scattering fields from multiple observing angles and frequencies need to be measured.

In this study, both transmitter and receiver work in "work-stop-work" mode which is applied to observe the vegetation scene as shown in Figure 2, and the field imaging algorithm suitable for this observing system is adopted to process the data. The UAV and satellite move in two straight lines parallel to the edge of the target area, respectively. The received signal of the UAV robot can be expressed as

$$E\left(\bar{k}\right) = \sum_{n=1}^{N} e^{i\bar{k}\cdot\vec{r}_n} F_n\left(\bar{k}\right), \quad (12)$$

where $\bar{k} = \bar{k}_i - \bar{k}_s$. $\bar{k}_i$ and $\bar{k}_s$ are the wave vectors of incident and scattered wave. N denotes the total number of scatterers. $F_n\left(\bar{k}\right)$ is the scattering amplitude of the nth scatterer. It is known that the received signal is an equation about the wave number and the receiving antenna position. Here, the orbit altitude of the transmitter of the transmitting radar remains unchanged.

In the actual calculation of the target function $F_n\left(\bar{k}\right)$, the received signal $E\left(\bar{k}\right)$ is not directly transformed by the inverse Fourier transform but by the conjugate multiplication and integration of received signal and the reference signal. The target of this study is virtual vegetation. The echo signal comes from the coherent scattering model of vegetation. The field imaging algorithm is to conjugate and multiply the echo signal at frequency k_m and position x_n with the reference signal, and then add and sum the echo signal at all frequencies and receiving antenna positions. Furthermore, the observation area is divided into $K \times L$ pixel units according to the requirement of imaging equality as shown in Figure 3. Each unit has a pixel center, which can be employed for the coordinate position of reference phase. The scattering intensity of pixel unit at $\bar{r}_0$ can be expressed as

$$E^s_{est}(\bar{r}_0) = \frac{1}{N_k N_x} \sum_{m=1}^{N_k} \sum_{n=1}^{N_x} E(k_m, x^t_n, x^r_n) E_0(k_m)^* e^{-ik_m(|\overline{R}_t(x^t_n) - \bar{r}_0| + |\overline{R}_r(x^r_n) - \bar{r}_0|)}, \quad (13)$$

where N_x and N_k represent the number of spatial sampling points of receiving antenna and discrete frequency points, respectively. The variable $\overline{R}_t(x^t_n) = x^t_n\hat{x} + y^t_n\hat{y} + z^t_n\hat{z}$ and $\overline{R}_r(x^r_n) = x^r_n\hat{x} + y^r_n\hat{y} + z^r_n\hat{z}$ represent the position vectors of transmitting and receiving antennas in the global coordinates. The position of the pixel center (k, l) is $\bar{r}_0 = x_k\hat{x} + y_l\hat{y} + 0\hat{z}$ in the global coordinates. $E_0(k_m)$ is the reference signal. $E(k_m, x^t_n, x^r_n)$ can be calculated by Equation (1). If we chose the different position $\bar{r}_0$ in Figure 3, Equation (13) can then be mapped into the imaging plane to reconstruct the multi-path scattering map.

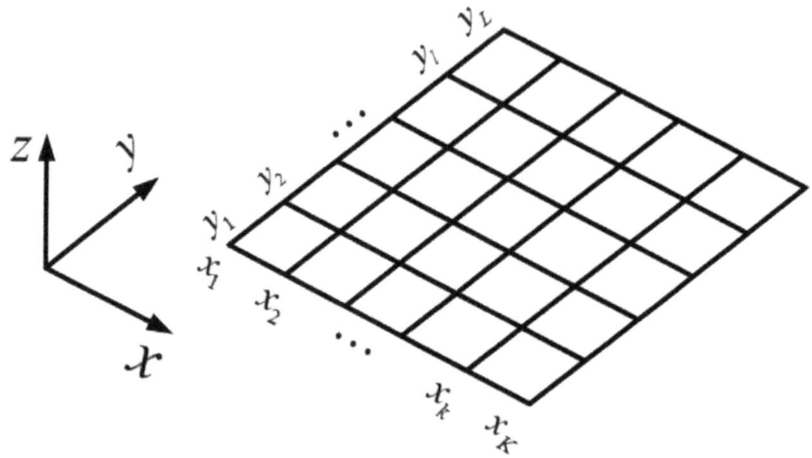

Figure 3. The grid division of imaging scene.

4. Results

In this simulation, we assume that the transmitter moves from −50 km to 50 km in the azimuthal direction, and that the receiver moves from −500 m to 500 m in the azimuthal direction. The frequency of signal gathering is from 6 GHz to 6.3 GHz with a step of 6 MHz. In the following, the visual vegetation is introduced firstly, and then the passive imaging results using multi-path scattering from single tree and multiple trees are presented.

4.1. Virtual Vegetation

In this study, L-system [19,20] is used to generate the 3-D spatial structures of trees, and the structure parameters of the tree are shown in Table 2. The dielectric constants of the tree elements and the ground surface at C band used in this simulation model are listed in Table 3.

Table 2. Structure parameters of trees.

Height of Trees		4 m	5 m	6 m
Leaf	Radius (m)	0.04	0.04	0.04
	Thickness (mm)	0.15	0.15	0.15
	Density (N m^{-3})	1360.0	580.0	250.0
Branch	Range of radius (cm)	0.1–1.6	0.1–2.2	0.1–2.7
	Range of length (cm)	1.0–103.8	1.0–178.8	1.0–195.0
	Density (N m^{-3})	180.0	60.0	45.0
Crown	Height (m)	2.0	3.0	3.2
	Width (m)	1.2	1.5	2.2

Table 3. The dielectric parameters of scatterers.

Parameters	Value
Leaves	20.24 + i6.78
Trunks	12.30 + i4.16
Branches	12.30 + i4.16
Ground	9.6 + i2.04

The ground surface is divided into many blocks according to the cell size used in calculating the canopy attenuations. The side length of the ground surface is 1 m. Moreover, the *rms* height and correlation length of the ground surface are set to be 1.5 cm and 15 cm, respectively.

4.2. The Passive Imaging Results Using Multi-Path Scattering from a Single Tree

Firstly, the simulation results of a single 5 m high tree located at the position (10 m, 10 m) in the coordinate system is presented and discussed. The height of the flight track of the transmitter and receiver remains unchanged, while the position of the receiver is changed to be at six different zenith angles. The model presented in Section 2 is used to simulate the electric field of this tree at different frequency points from 6 GHz to 6.3 GHz, with a step of 6 MHz and a position from −500 m to 500 m with a step of 5 m. Next, the simulating electric fields received by UAV robot are processed by the imaging method described in Section 3 to obtain the images. The following six images are obtained by above methods at zenith angles of 26°, 36°, 46°, 56°, 66° and 76°, respectively, as shown in Figure 4a–f. The scene size is 20 m × 20 m, and the size of each pixel is 0.2 m × 0.2 m.

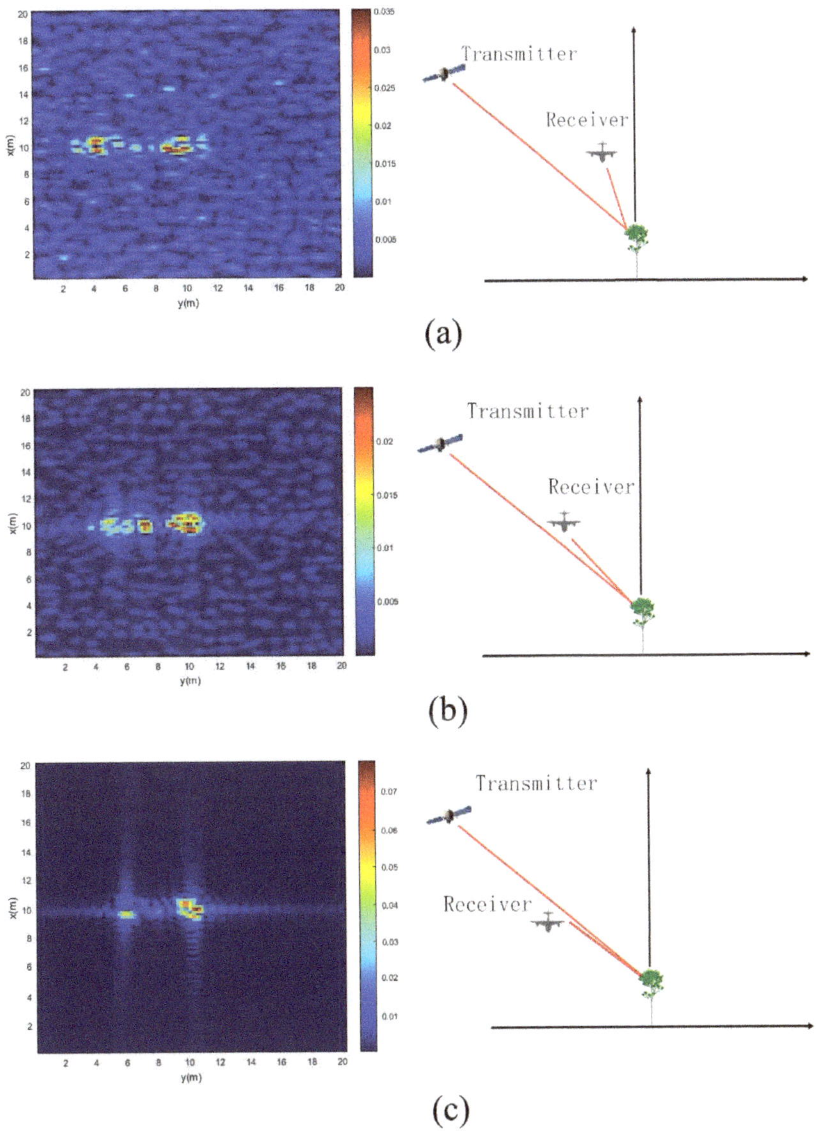

Figure 4. *Cont.*

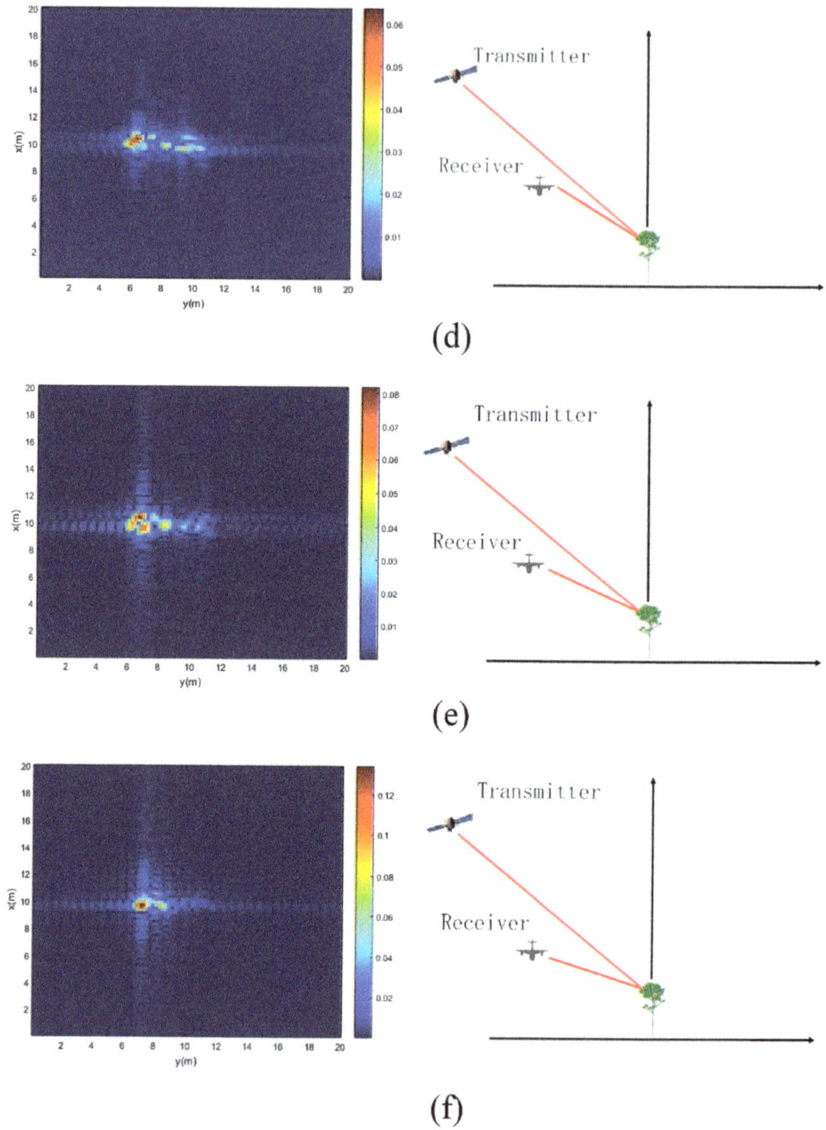

Figure 4. UAV imaging results of a single tree at different receiver zenith angles: (**a**) receiving zenith angle = 26°; (**b**) receiving zenith angle = 36°; (**c**) receiving zenith angle = 46°; (**d**) receiving zenith angle = 56°; (**e**) receiving zenith angle = 66°; (**f**) receiving zenith angle = 76°.

From the images, it can be seen that the simulation results demonstrate good performance of the autofocus based on the passive imaging algorithm using multi-path scattering. The energy concentrates on two main parts; the left part is mainly from the direct scattering of the scatterer in the vegetation, and the right part is the summation of a single bounce between the scatterer and the ground. The double bounce of ground-scatterer-ground is too weak to be shown in the images.

4.3. The Passive Imaging Results Using Multi-Path Scattering from Multiple Trees

Furthermore, multiple trees in an area of 30 m × 30 m are generated by the L-system. The distributions and heights of trees in this area are shown in Figure 5.

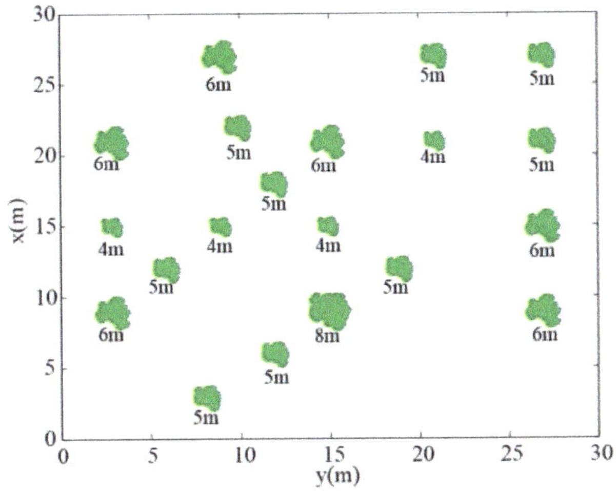

Figure 5. The distribution and height of each tree in a stand.

Figure 6 shows the bistatic UAV microwave imaging results of multiple of trees in this stand. The height of flight track of the transmitter and receiver also remains unchanged, while the position of the receiver is changed to be at six different zenith angles. The rest of the bistatic UAV microwave imaging parameters are the same as those in Section 2.2. Six imaging results of multiple trees are obtained at zenith angles of 26°, 36°, 46°, 56°, 66° and 76°, respectively, as shown in Figure 6a–f.

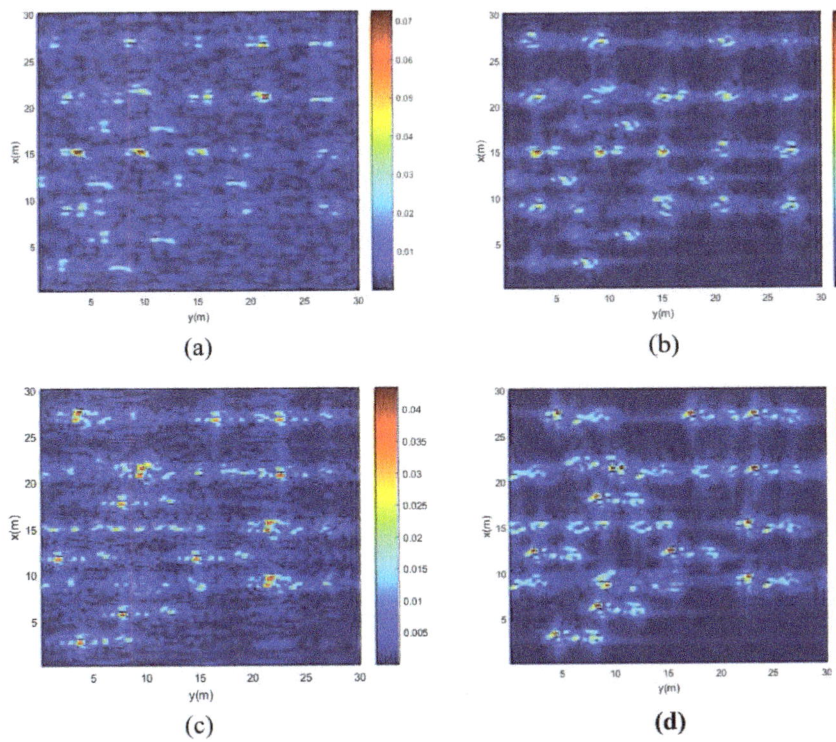

Figure 6. *Cont.*

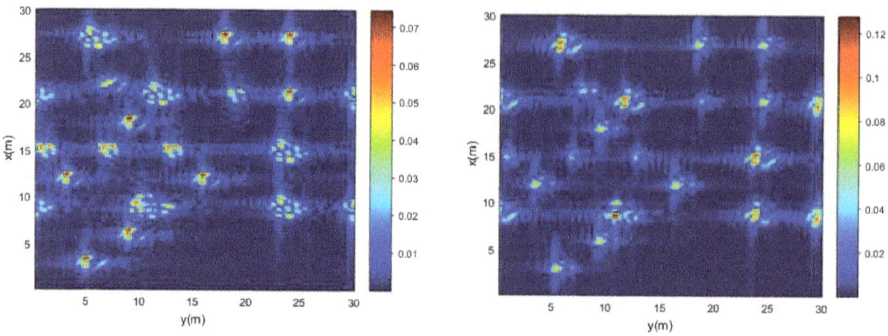

Figure 6. UAV imaging results of trees at different receiver zenith angles: (**a**) receiving zenith angle = 26°; (**b**) receiving zenith angle = 36°; (**c**) receiving zenith angle = 46°; (**d**) receiving zenith angle = 56°; (**e**) receiving zenith angle = 66°; (**f**) receiving zenith angle = 76°.

5. Discussion

From above simulation results, based on the imaging method proposed in this study, it can be seen that the larger the receiver's zenith angle is, the closer the distance between two parts of energy becomes. This is because the distributions of scattering energy in these images are related to the travel path and the imaging algorithm. When the receiver's zenith angle is larger, the lengths of propagation paths from direct scattering and single bounce become closer. For this imaging configuration, it can be inferred that when the receiver's zenith angle becomes large enough, the two parts will overlap.

In addition, it can be observed that the range resolution of Figure 6c–f is relatively clear among the six images. The resolution of the image shown in Figure 6b is somewhat worse than that in Figure 6c, but better than that in Figure 6a. These results might be due to the variation in the receiver's zenith angle. Equation (14) presents the range resolution formula [21], where c is the velocity of an electromagnetic wave propagating in free space, and B is bandwidth. θ_t and θ_r are the zenith angles of transmitter and receiver, respectively. The range resolutions as a function of zenith angle of receiver are calculated based on Equation (14), as shown in Figure 7. The range resolutions of six zenith angles of receivers shown in Figure 6 are listed in Table 4. The calculation results contribute to the understanding of the aforementioned imaging results.

$$\Delta L_{range} = \frac{c}{B(\sin(\theta_t) + \sin(\theta_r))} \tag{14}$$

Table 4. Range resolution of different receiving angles.

Zenith Angle of Receiver (°)	Value (m)
26	0.87
36	0.77
46	0.70
56	0.65
66	0.62
76	0.60

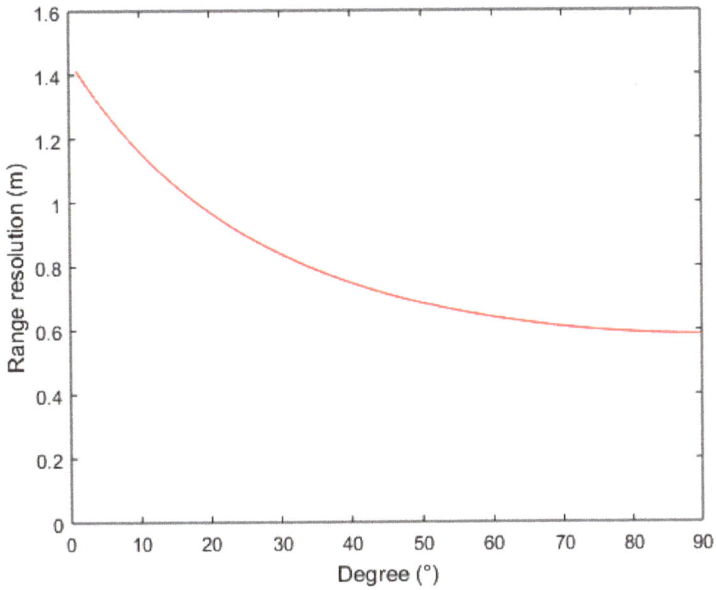

Figure 7. The range resolutions as a function of the zenith angle of a receiver.

Furthermore, from Figure 6, it is noticed that the zenith angle of the receiver also affects the signal strength received from the ground surface. Direct scattering from the ground surface is clearly shown in Figure 6a. However, as the zenith angle of the receiver increases, the signal strength from the ground surface becomes weaker and weaker. For Figure 6c, the signal strength received from the ground surface is too weak to be distinguished. In addition, it can be seen from the figures that the scattering strength of a single tree from the left part is gradually stronger than that on the right part. We have suggested in Section 4.2 that the image of a single tree includes two main parts: the left part is mainly from direct scattering of the scatterer in the tree, and the right part is the summation of a single bounce between the scatterer and ground. For a large receiving angle, the signal strength received from the ground surface is relatively weak. Furthermore, the strength from right part in the figure will also weaken because the right part depends on a single bounce between the scatterer and the ground. Therefore, it can be seen that the signal strength from the left part becomes gradually stronger, as shown in Figure 6a–f.

For imaging results of multiple trees in a stand as shown in Figure 6, it can be found that when the receiver zenith angle is large, the distribution of trees in this area is relatively clear, and different scattering components of a single tree are relatively close. When the receiver zenith angle is small, the scattering component received from the ground surface is strong. To a certain extent, some scattering components of the tree are suppressed by direct scattering components of the ground surface, as shown in Figure 6a. In Figure 6a, we cannot clearly distinguish the complete scattering information of every tree, we can only roughly distinguish the position of every tree from the image.

6. Conclusions

This article presents the simulation results of UAV robot passive imaging using the satellite communication systems based on the multi-path scattering model. In this UAV robot imaging configuration, the incident angle of the transmitter remains unchanged, while the different flight tracks of receivers loaded on different UAVs result in different zenith angles. Furthermore, the simulated signals received from the multi-path scattering model at each position are processed by a passive imaging algorithm. From the imaging results of vegetation, we can see that different zenith angles of the receiver affect the imaging results. When the receiving and transmitting antenna is located at the same side of the target area, the greater the zenith angle of the receiving antenna is, the higher the range resolution of the UAV robot passive imaging is. The combination of multi-path scattering model and the imaging algorithm provides an effective method to study the scattering mechanism based on vegetation. At the same time, the method can be utilized to study the effects of vegetation scattering. The simulation results offer a reference for the development and research of UAV robot passive imaging using the satellite communication systems in the field of ground microwave monitoring.

Author Contributions: T.H. conceived the idea and developed multi-path scattering model. Z.C. and X.Q. modified the model. X.Q. performed numerical simulation and the passive imaging calculations. Z.C. and T.Z. wrote the manuscript, and T.H. revised the manuscript. P.W. and L.F. analyzed the simulation results. All authors have read and agreed to the published version of the manuscript.

Funding: This research was funded by Southwest Technology and Engineering Research Institute of China Cooperation fund grant number HDHDW5901010301.

Institutional Review Board Statement: Not applicable.

Informed Consent Statement: Not applicable.

Data Availability Statement: Data available on request from the authors.

Conflicts of Interest: The authors declare that they have no conflict of interest.

References

1. Li, H.; Savki, A.V. Wireless sensor network based navigation of micro flying robots in the industrial internet of things. *IEEE Trans. Ind. Inform.* **2018**, *14*, 3524–3533. [CrossRef]
2. Jang, D.; Yoo, J.; Son, C.Y.; Kim, D.; Kim, H.J. Multi-Robot Active sensing and environmental model learning with distributed gaussian process. *IEEE Robot. Autom. Lett.* **2020**, *5*, 5905–5912. [CrossRef]
3. Yoo, S.; Lee, W. Federated Reinforcement Learning Based AANs with LEO Satellites and UAVs. *Sensors* **2021**, *21*, 8111. [CrossRef] [PubMed]
4. Huang, Y.; Liu, F.; Chen, Z.; Li, J.; Hong, W. An improved map-drift algorithm for unmanned aerial vehicle SAR imaging. *IEEE Geosci. Remote Sens. Lett.* **2021**, *18*, 1–5. [CrossRef]
5. Bouchat, J.; Tronquo, E.; Orban, A.; Verhoest, N.E.C.; Defourny, P. Assessing the potential of fully polarimetric mono- and bistatic sar acquisitions in L-Band for crop and soil monitoring. *IEEE J. Sel. Top. Appl. Earth Observ. Remote Sens.* **2022**, *15*, 3168–3178. [CrossRef]
6. Pierdicca, N.; Brogioni, M.; Fascetti, F.; Ouellette, J.D.; Guerriero, L. Retrieval of biogeophysical parameters from bistatic observations of land at L-band: A theoretical study. *IEEE Trans. Geosci. Remote Sens.* **2021**, *60*, 4402517. [CrossRef]
7. Tang, S.; Guo, P.; Zhang, L.; Lin, C. Modeling and precise processing for spaceborne transmitter/missile-borne receiver SAR signals. *Remote Sens.* **2019**, *11*, 346–363. [CrossRef]
8. Zhou, X.; Chen, J.; Wang, P.; Zeng, H.; Fang, Y.; Men, Z.; Liu, W. An efficient imaging algorithm for GNSS-R Bi-Static SAR. *Remote Sens.* **2019**, *11*, 2945–2968. [CrossRef]
9. Wang, Z.; Liu, F.; Zeng, T.; Wang, C.A. Novel motion compensation algorithm based on motion sensitivity analysis for Mini-UAV-based BiSAR system. *IEEE Trans. Geosci. Remote Sens.* **2021**, *60*, 5205813. [CrossRef]
10. Li, Z.; Zhang, X.; Yang, Q.; Xiao, Y.; An, H.; Yang, H.; Wu, L.; Yang, J. Hybrid SAR-ISAR image formation via joint FrFT-WVD processing for BFSAR ship target high-resolution imaging. *IEEE Trans. Geosci. Remote Sens.* **2021**, *60*, 5215713. [CrossRef]
11. Focsa, A.; Anghel, A.; Datcu, M.A. Compressive-sensing approach for opportunistic bistatic SAR imaging enhancement by harnessing sparse multiaperture data. *IEEE Trans. Geosci. Remote Sens.* **2021**, *60*, 5205914. [CrossRef]
12. Guerriero, L.; Pierdicca, N.; Pulvirenti, L.; Ferrazzoli, P. Use of satellite radar bistatic measurements for crop monitoring: A simulation study on corn fields. *Remote Sens.* **2013**, *5*, 864–890. [CrossRef]
13. Bellez, S.; Roussel, H.; Dahon, C.; Castelli, J.C.; Cheraly, A. Full polarimetric bistatic radar imaging experiments on sets of dielectric cylinders above a conductive circular plate. *IEEE Trans. Geosci. Remote Sens.* **2013**, *51*, 4164–4176. [CrossRef]
14. Yang, H.; Liu, D.; Sun, G.; Guo, Z.; Zhang, Z. Simulation of interferometric SAR response for characterizing forest successional dynamics. *IEEE Geosci. Remote Sens. Lett.* **2014**, *11*, 1529–1533. [CrossRef]
15. Tsang, L.; Kong, J.A.; Shin, R.T. *Theory of Microwave Remote Sensing*; Wiley-Interscience: New York, NY, USA, 1985.
16. Fung, A.K.; Chen, K.S. *Microwave Scattering and Emission Models and Their Applications*; Artech House: Norwood, MA, USA, 1994.
17. Liu, D.; Sun, G.; Guo, Z.; Ranson, K.J.; Du, Y. Three-dimensional coherent radar backscatter model and simulations of scattering phase center of forest canopies. *IEEE Trans. Geosci. Remote Sens.* **2010**, *48*, 349–357.
18. Tsang, L.; Kong, J.A. *Scattering of Electromagnetic Waves: Numerical Simulation*; Wiley-Interscience: New York, NY, USA, 2001.

19. Lindenmayer, A. Mathematical models for cellular interactions in development I. Filaments with 1-sided inputs. *J. Theor. Biol.* **1968**, *18*, 280–299. [CrossRef]
20. Prusinkiewicz, P.; Lindenmayer, A. *The Algorithmic Beauty of Plants*; Springer: New York, NY, USA, 1990.
21. Sanz-Marcos, J.; Lopez-Dekker, P.; Mallorqui, J.J.; Aguasca, A.; Prats, P. SABRINA: A SAR bistatic receiver for interferometric applications. *IEEE Geosci. Remote Sens. Lett.* **2007**, *4*, 307–311. [CrossRef]

sensors

MDPI

Communication

A Reinforcement Learning-Based Strategy of Path Following for Snake Robots with an Onboard Camera

Lixing Liu, Xian Guo * and Yongchun Fang

Institute of Robotics and Automatic Information System, College of Artificial Intelligence, Nankai University, Tianjin 300071, China
* Correspondence: guoxian@nankai.edu.cn

Abstract: For path following of snake robots, many model-based controllers have demonstrated strong tracking abilities. However, a satisfactory performance often relies on precise modelling and simplified assumptions. In addition, visual perception is also essential for autonomous closed-loop control, which renders the path following of snake robots even more challenging. Hence, a novel reinforcement learning-based hierarchical control framework is designed to enable a snake robot with an onboard camera to realize autonomous self-localization and path following. Specifically, firstly, a path following policy is trained in a hierarchical manner, in which the RL algorithm and gait knowledge are well combined. On this basis, the training efficiency is sufficiently optimized, and the path following performance of the control policy is greatly improved, which can then be implemented on a practical snake robot without any additional training. Subsequently, in order to promote visual self-localization during path following, a visual localization stabilization item is added to the reward function that trains the path following strategy, which endows a snake robot with smooth steering ability during locomotion, thereby guaranteeing the accuracy of visual localization and facilitating practical applications. Comparative simulations and experimental results are illustrated to exhibit the superior performance of the proposed hierarchical path following the control method in terms of convergence speed and tracking accuracy.

Keywords: snake robots; visual localization; path following; reinforcement learning control

Citation: Liu, L.; Guo, X.; Fang, Y. A Reinforcement Learning-Based Strategy of Path Following for Snake Robots with an Onboard Camera. *Sensors* **2022**, *22*, 9867. https://doi.org/10.3390/s22249867

Academic Editor: Sašo Blažič

Received: 14 November 2022
Accepted: 12 December 2022
Published: 15 December 2022

1. Introduction

Over the past decades, many researchers have been devoted to the control of snake robots due to their complex multi-joint structure and high motion flexibility [1–3]. As the application tasks of snake robots become more complex, the requirement for accomplishing a safe and accurate path following tasks with independent perception continues to increase. Specifically, path following, as one of the most fundamental and indispensable motion skills, requires the robot to move along a specific curve. Visual self-localization provides the real-time position of the robot via visual perception, which plays an important role in assisting a robot in completing autonomous motions. However, the highly redundant degrees of freedom and unique serpentine motion gait introduce many challenges to the path following of snake robots with visual self-localization.

Snake robots typically move forward by mimicking the motion gait of biological snakes. One of the most efficient and widely used motion gaits is the lateral undulatory gait; that is, it periodically propagates a wave along the body, presenting an S-shaped movement trajectory, which is named a serpenoid by Hirose [1]. For snake robots, the lateral undulatory gait can be mathematically approximated by a gait equation that imposes a sinusoidal signal for each joint. The path following control of snake robots has been investigated for years based on the gait equation. There is substantial research that focuses on designing a control law for the gait equation to adjust the motion direction and thus control the robot in moving towards the desired path with the desired turning angle, which is calculated by the line-of-sight (LOS) guidance law [4–6]. Furthermore, for more complicated application scenarios, path following controllers with an adaptive LOS guidance law or gait equation are adopted for faster convergence speed, and higher stability [7–9]. In addition, for holonomic snake robots, virtual constraints are employed to regulate the orientation, and forward speed of the snake robot via a state-dependent undulatory gait equation, which replaces time-dependent signals in the lateral undulatory gait equation with state-dependent constraints [10–13]. However, the methods mentioned above heavily rely on precise modelling and laborious parameter tuning, which cannot guarantee optimal path following performance.

Reinforcement learning (RL), as a popular machine learning algorithm that constantly interacts with the environment to maximize expected returns, has made much progress in robotic control [14–16]. Unlike traditional control methods, RL algorithms endow robots with various motion skills without acquiring the exact robot model and exhibit excellent robustness and flexibility regarding environment variations. In addition, in RL, the control objectives and constraints can be conveniently added as terms to the reward function to guide robots to complete the specified task, e.g., manipulator manipulation, tracking a target velocity, mimicking human motor skills, etc. Model-free (MF) RL has shown its unique advantages in mastering specific skills or accomplishing specific tasks. However, end-to-end MFRL training often suffers from sample inefficiency and is prone to generating weird and unnatural actions, which seriously reduce training speed, and the learned policy may even damage the practical robot. Consequently, it is essential to integrate RL algorithms with gait knowledge to improve training efficiency and generate natural actions that make deployment on a practical robot easier. Recently, RL algorithms assisted by gait knowledge have made great progress in the fields of quadruped robots, bipedal robots, etc. [17–21]. For snake robots, the gait equation can reflect the shape of the motion trajectory; thus, it is an ideal source of gait knowledge. However, incorporating the gait equation with the RL algorithm for path following tasks is still a difficult problem. In our prior work [22], a two-stage control framework that combines PI^2 with the gait equation is proposed for the snake robot to perform goal-driven tasks, but the gait equation is only adjusted at each gait cycle. As a result, the robot cannot promptly correct its gait according to the position error, so the control accuracy of this method is too low for performing path following tasks.

In addition to path following control, satisfactory visual self-localization of snake robots is also a challenging topic. Due to the slender body structure and serpentine motion gait, a snake robot obtains forward momentum using whole-body motion; thus, the camera installed on the robot also shakes accordingly, which brings difficulties to stable visual perception and localization. To solve this problem, the robot is usually required to remain stationary or to move slowly during the imaging process, which optimizes localization accuracy but sacrifices locomotion efficiency [23,24]. Ref. [25] proposes a pan-tilt compensation strategy to realize visual self-localization without reducing the robot's locomotion efficiency, in which the position of the robot is updated by an external visual marker. The camera is mounted on a pan-tilt that actively rotates to compensate for head motions, which ensures that the visual marker is always within the camera's field of view. Consequently, this method requires relatively smooth head swings at adjacent moments to guarantee the successful compensation of the pan-tilt. However, during the training of the path following policy using RL algorithms, a snake robot is prone to learning a policy that gains a higher reward for approaching the desired path faster, leading to an aggressive policy; i.e., it vigorously wiggles the entire body from side to side to gain stronger forward momentum, based on which, the pan-tilt cannot promptly compensate for the motion of the head; thus, the camera installed on the pan-tilt will lose the target marker.

To solve the above-mentioned problems, a novel hierarchical path following the control method is proposed for snake robots, which presents high training efficiency and promising tracking performance. Specifically, the hierarchical path following the control method is developed by combining the RL algorithm and the gait equation. On the one hand, the hierarchical control method generates motion gaits by modifying the gait equation, which provides gait knowledge for the RL algorithm and thus accelerates the training process. In addition, it ensures that the resulting motor gait belongs to the lateral undulatory gait so that the learned policy can be directly transferred to the practical snake robot without any retraining. On the other hand, the proposed hierarchical method adjusts the gait equation in real-time via the RL algorithm, which expands the feasible gait set of the gait equation, thus enabling a snake robot to change the motion gait based on the feedback state promptly. In addition, in order to enable a snake robot to achieve visual self-localization using the pan-tilt compensation strategy, a visual localization stabilization item is added to the reward function of RL policy training, which effectively limits the swing amplitude of the head at adjacent moments. The contributions of this paper are summarized as follows:

- A novel hierarchical control method that combines the RL algorithm and the gait equation is developed for the path following of snake robots, which guarantees efficient training and is satisfactory following the accuracy.
- A visual localization stabilization term is introduced into the reward function to avoid excessive head swings, which ensures successful pan-tilt compensation, thereby optimizing the accuracy of visual localization.
- To verify the effectiveness of the algorithm, real-world experiments are implemented on a practical snake robot, and the experimental results demonstrate the promising path following the performance of the proposed method.

2. Materials and Methods

2.1. Problem Statement

The path-following task of snake robots can be formulated as a Markov decision process (MDP) defined by the tuple $(\mathcal{S}, \mathcal{A}, \mathcal{R}, \mathcal{P}, \gamma)$, where $\mathcal{S}$ denotes the state space, $\mathcal{A}$ represents the action space, $\mathcal{R}$ is the reward function, $\mathcal{P}$ indicates the state transition probability, and γ stands for the discount factor. At each timestep, the snake robot samples and then executes the action $\boldsymbol{a}_t \sim \pi(\boldsymbol{a}_t \mid \boldsymbol{s}_t)$ based on current state $\boldsymbol{s}_t$ observed from the environment, and then the robot transfers to a new state $\mathbf{s}_{t+1} \sim p(\mathbf{s}_{t+1} \mid \mathbf{s}_t, \mathbf{a}_t)$ and receives a reward $r_t = r(\mathbf{s}_t, \mathbf{a}_t, \mathbf{s}_{t+1})$, where the subscript $\star_t$ depicts the current timestep t. The objective of this MDP is to train a policy π^*_ψ with parameter ψ that maximizes the expected cumulative discounted return $\boldsymbol{J}$, as indicated in Equations (1) and (2), so as to equip a snake robot with the excellent path following skill:

$$\pi^*_\psi = \arg\max_{a \in \mathcal{A}} \boldsymbol{J} \tag{1}$$

$$\boldsymbol{J} = \mathbb{E}_{\tau \sim p(\tau|\pi)} \left[\sum_{t=0}^{T-1} \gamma^t r_t \right] \tag{2}$$

where T denotes the planning horizon of each episode, and τ denotes a trajectory $\{\mathbf{s}_0, \mathbf{a}_0, r_0, \mathbf{s}_1, \ldots, \mathbf{s}_{T-1}, \mathbf{a}_{T-1}, r_{T-1}, \mathbf{s}_T\}$.

2.2. Hierarchical Path Following Control

In this paper, the objective is to design a controller that enables a snake robot to follow the given path with visual self-localization. To this end, we proposed a hierarchical RL path following method to guarantee satisfactory following the ability for various desired paths in terms of efficient training, strong robustness, and excellent following accuracy. The proposed hierarchical algorithm effectively incorporates the RL algorithm with the gait equation and consists of two layers, namely the RL policy training layer and the gait execution layer. Specifically, compared with the motion gait produced by the traditional gait equation, the proposed method tends to generate a forward gait with slighter head swings, which improves the accuracy of visual localization and further guarantees satisfaction following accuracy. In addition, compared with the end-to-end RL algorithm, the designed hierarchical RL method not only greatly accelerates the training speed but also learns a natural and robust policy that can be directly implemented on a practical snake robot.

The overall architecture of the proposed control method is presented in Figure 1, which consists of two stages, namely the visual localization and hierarchical RL path following policy training. Specifically, for a $n-$link snake robot, at each timestep t, firstly, the pan-tilt compensate strategy proposed in [25] is adapted to provide the real-time position of a snake robot, which can be used to obtain the current system state; subsequently, the RL policy training layer outputs an action to modify the gait parameter of the gait equation based on the state, with the aim of changing the motion direction of the robot to make it close to the desired path. Finally, the gait execution layer sends the corresponding joint angles to the snake robot for executing the path following task.

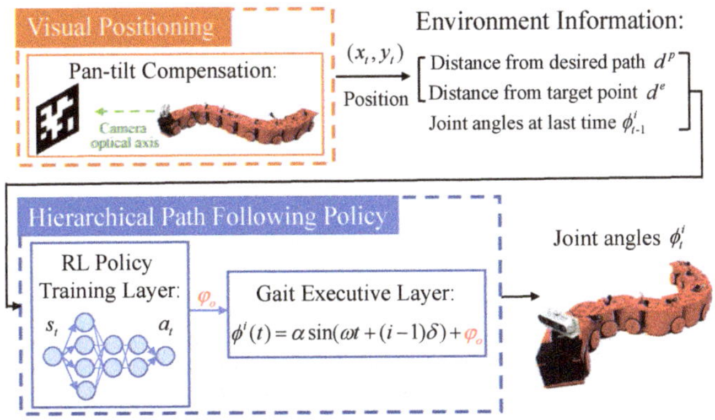

Figure 1. The overall control architecture of the proposed path following method, which consists of two stages: visual self-localization and hierarchical path following control.

2.2.1. Visual Localization

Due to the head swings caused by the lateral undulatory gait, the camera mounted on the robot head always loses the visual marker, so the position of the robot cannot be updated in real-time. Therefore, Ref. [25] proposes a pan-tilt compensation strategy to always keep the camera plane parallel to the visual marker plane via active compensation, where the compensation angle of the pan-tilt θ_t^{PT} is represented as follows:

$$\theta_t^{PT} = -\theta_t^{head} \tag{3}$$

where θ_t^{head} is the orientation angle of the head of a snake robot at timestep t, and the compensation angle θ_t^{PT} is only related to θ_t^{head} with the same value and the opposite direction; that is, if the head turns to the left, the pan-tilt automatically rotates the same angle to the right to keep the camera facing the visual target. θ_t^{head} can be indicated in the following manner:

$$\theta_t^{head} = \frac{\pi}{2} - \theta_t^{ac} - \theta_{t-1}^{PT} \tag{4}$$

where θ_t^{ac} denotes the current deviation angle between the visual marker plane and the camera plane after the last pan-tilt compensation. After the pan-tilt compensation, the position of the head of a snake robot can be calculated through visual localization and coordinate transformation as follows:

$$\begin{aligned} X_{cam}^{w} &= R_{tgt}^{w} X_{cam}^{tgt} + p_{tgt}^{w} \\ X_{head}^{w} &= R_{cam}^{w} X_{head}^{cam} + X_{cam}^{w} \end{aligned} \tag{5}$$

where X_{cam}^{w} is the coordinate of the camera mounted on the head, which can be calculated by rotation matrix R_{tgt}^{w} and translation matrix p_{tgt}^{w} from the visual target coordinate system to the world coordinate system and the position of camera X_{cam}^{tgt} in the visual target coordinate system. Furthermore, X_{head}^{w} is the position of the head of the snake robot, R_{cam}^{w} denotes the rotation matrix between the camera coordinate system and the world coordinate system, and X_{head}^{cam} expresses the coordinate of the head in the camera coordinate system.

Based on the above introduction, it can be derived that the angle of the robot head plays an important role in determining the accuracy of the visual localization. Excessive head swings lead to loss of visual markers, which further results in the failure of path-tilt compensation. Consequently, a visual localization stabilization term is proposed to reduce head swings during motion and embedded in the training process of the path following strategy, which will be described in detail below.

2.2.2. Rl Policy Training Layer

As the first stage of the hierarchical control framework, the training objective of the RL policy training layer is to find a policy that outputs the optimal action according to the current state of the whole system, which will be used to modify the gait equation [1] in the later control stage to ensure good path following performance. To this end, the high-level control law of the hierarchical control framework is defined as follows:

$$u_{high} = \boldsymbol{a}_t \sim \pi_{\psi}(\boldsymbol{a}_t \mid \boldsymbol{s}_t) \tag{6}$$

In order to achieve satisfactory path following performance, a snake robot is expected to approach the desired path as close as possible; in addition, to prevent the robot from stopping as soon as it approaches the path, random target points are selected on the desired path along the forward direction of the robot to guide its forward motion while continuing the path following.

The state of the path following task is embedded in vector $\boldsymbol{s}_t \in \mathbb{R}^{n+1}$, which consists of the distance between the real-time position of the snake robot and the desired path $\boldsymbol{d}^p$, the distance between the robot and the endpoint d^e, and the joint angles command $\boldsymbol{\phi}_{t-1}^{i}, i = 1, 2, ..., n-1$ at the last timestep.

According to state $\boldsymbol{s}_t$, action $\boldsymbol{a}_t \in \mathbb{R}^1$ drives the snake robot to follow the desired path by adjusting the parameter of the gait equation, which generates the lateral undulatory gait as the Equation (7), with α, ω, δ, and ϕ_o denoting the gait amplitude, angular frequency, phase difference, and offseting of the lateral undulatory gait, respectively.

$$\boldsymbol{\phi}^{i}(t) = \alpha \sin(\omega t + (i-1)\delta) + \phi_o \tag{7}$$

where $\boldsymbol{\phi}^{i}(t), i = 1, 2, ..., n-1$ is the i-th joint angle of the snake robot at time t. Different groups of four gait parameters generate different forms of the motion trajectory of the lateral undulatory gait. To drive the robot to follow the desired path, the joint offset ϕ_o, which can modify the motion direction in real-time, is selected as the action and then generated by the policy network.

$$\boldsymbol{a}_t = \phi_o \tag{8}$$

To improve the accuracy and efficiency of the path following, the reward function is designed as follows:

$$r_t = r_p + r_e - p_h \tag{9}$$

where r_p encourages the snake robot to approach the desired path with a defined tolerance, the second term r_e rewards the robot for moving forward towards the endpoint as soon as possible, and the last term p_h is the visual localization stabilization term, which penalizes the robot for excessive head swings in adjacent moments. Specifically, the three terms are constructed as follows:

$$r_p = \begin{cases} c_p & \text{if } \left|\boldsymbol{d}_{t+1}^{p}\right| < d_1 \\ c_p \exp\left(d_1 - \left|\boldsymbol{d}_{t+1}^{p}\right|\right) & \text{if } d_1 \leq \left|\boldsymbol{d}_{t+1}^{p}\right| \leq d_2 \\ 0.0 & \text{if } \left|\boldsymbol{d}_{t+1}^{p}\right| > d_2 \end{cases} \tag{10}$$

$$r_e = c_e(d_t^e - d_{t+1}^e) \tag{11}$$

where c_p and c_e are the weighting constants, and d_1 and d_2 are the distance thresholds at which the reward approaching the goal path can be obtained.

In order to improve the accuracy of the visual localization mentioned in Section 2.2.1, the visual localization stabilization term p_h is depicted as follows, with c_h being a negative constant and ϕ_* being the angle threshold:

$$p_h = \begin{cases} c_h & \text{if } \left|\boldsymbol{\phi}_{t+1}^1 - \boldsymbol{\phi}_t^1\right| \geq \phi_* \\ 0.0 & \text{if } \left|\boldsymbol{\phi}_{t+1}^1 - \boldsymbol{\phi}_t^1\right| < \phi_* \end{cases} \tag{12}$$

where $\boldsymbol{\phi}_t^1$ and $\boldsymbol{\phi}_{t+1}^1$ are the head angles of the snake robot at timestep t and $t+1$.

The Proximal Policy Optimization (PPO) algorithm is adopted to train policy π_ψ, which is represented by a fully connected network with 3 Tanh hidden layers of $[64, 32, 32]$ units.

2.2.3. Gait Executive Layer

The gait executive layer is the second stage of the hierarchical control framework, which is controlled by the high-level action $\boldsymbol{a}_t$ and presents a modified motion gait via the gait equation shown in Equation (7). The low-level control law is illustrated as follows:

$$u_{low} = \phi^i(t) \tag{13}$$

The generated control command u_{low} of the joint angels is directly sent to a snake robot and helps it to change the motion direction and then converge to the desired path.

3. Results

In this section, the hierarchical path following policy is firstly trained and then tested in the simulation, and the training efficiency and the effectiveness of the proposed algorithm are verified. Subsequently, the trained policy is directly transferred to real-world experiments, and several experiments are implemented on a practical snake robot to validate the actual following performance for different desired paths.

3.1. Simulations

The simulation environment is developed based on the Mujoco [26] simulator, and the model of the snake robot is composed of nine connection modules with a pair of passive wheels and eight yaw joints. During training, the start point of the snake robot is $(0\,\text{m}, 0\,\text{m})$. The desired paths are straight lines $y = y^* \in [-1.5\,\text{m}, 1.5\,\text{m}]$, sinusoidal curves $y = A \sin \omega x + \phi$, $A \in [0.2\,\text{m}, 1.0\,\text{m}]$, $\omega \in [\frac{\pi}{2}\text{rad}, \pi\text{rad}]$, $\phi \in [-1.5\,\text{m}, 1.5\text{m}]$, and circles $x^2 + y^2 = R^2, R \in [1.5\,\text{m}, 3.0\,\text{m}]$, respectively. The target point is a random point on the desired path with an $x-$coordinate $x^* \in [4.0\,\text{m}, 5.0\,\text{m}]$. The end-to-end PPO algorithm, which takes the same state as the input and directly outputs joint angles, is selected as the comparative method to demonstrate the training efficiency and tracking performance of the proposed algorithm. Firstly, the comparison of the mean reward of an episode between the two methods is depicted in Figure 2.

The training results demonstrate that the proposed hierarchical control method achieves a superior training performance with a higher episode reward and faster convergence speed compared with the comparative method. Specifically, the proposed method converges to higher episode reward within about 1M timesteps, while the comparative method slowly converges to a reward value that is less than one-third of that of the proposed method at about 2M timesteps, which clearly indicates the strong path following ability and reliable training efficiency of the proposed method.

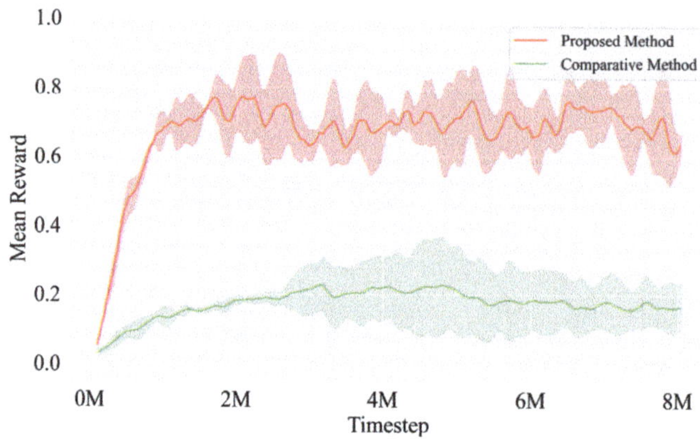

Figure 2. Mean reward of an episode for the proposed hierarchical method and the end-to-end comparative method.

To verify the following performance of the learned policy, the path following tests are performed on three different types of desired paths, and the following results are shown in Figures 3–5. It is indicated that the snake robot converges to the desired path agilely and accurately, and then it keeps following the path with small tracking errors driven by the proposed method, implying the superior path following ability of the hierarchical trained policy; in comparison, under the control of the end-to-end comparative method, the snake robot requires longer convergence time and presents larger tracking errors.

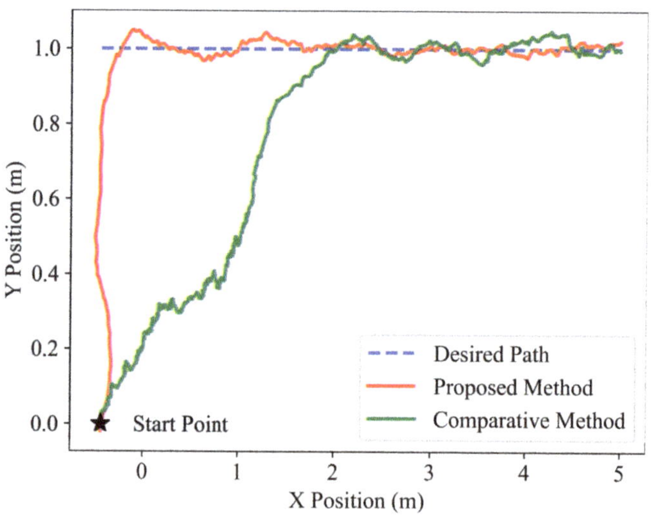

Figure 3. Simulation path following results of the desired line for the proposed hierarchical method and the end-to-end comparative method.

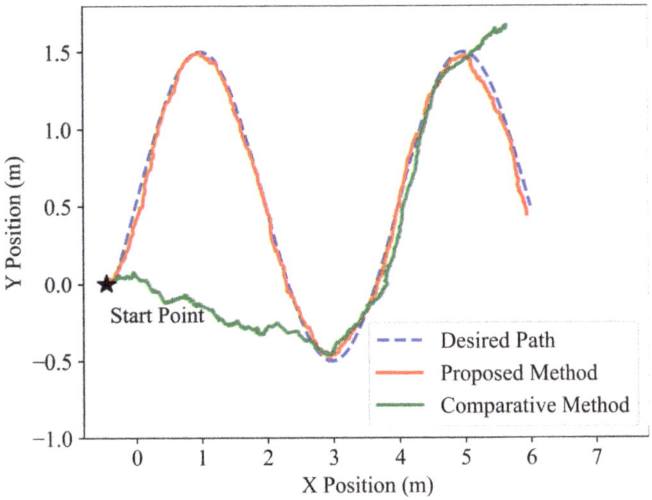

Figure 4. Simulation path following results of the desired sinusoidal curve for the proposed hierarchical method and the end-to-end comparative method.

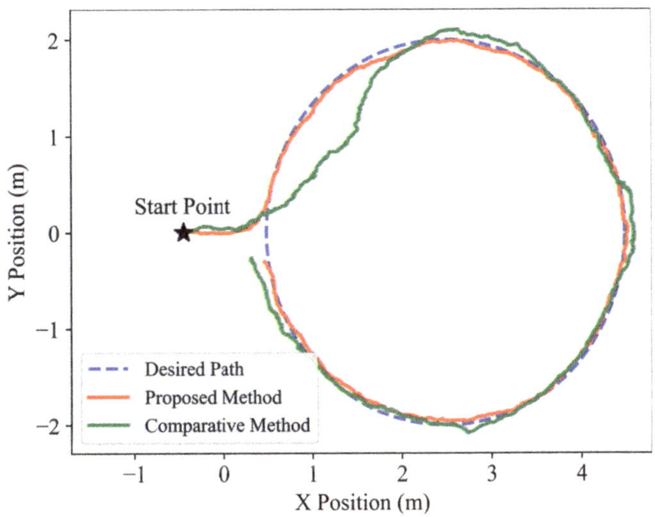

Figure 5. Simulation path following results of the desired circle for the proposed hierarchical method and the end-to-end comparative method.

3.2. Experiments

To validate the actual performance of the proposed method, two groups of hardware experiments are conducted on a self-built practical snake robot, and the results and analyses are provided in this part. Specifically, the snake robot is composed of visual localization module and motion modules, where the visual localization module contains a RealSense D435i camera for capturing images and a Dynamixel AX-12A actuator serving as the pan-tilt, and each 3D-printed body module has a mass of 0.416 g; it consists of a Hitec HS-5585MH actuator, a lithium battery pack with a voltage of 7.4 V, a wireless serial port and a pair of passive wheels.

Firstly, to examine whether visual localization stabilization term p_h improves the localization accuracy by reducing head swings, we compare the accuracy of the visual localization for the trained policies with and without the stabilization item p_h, and the comparative results are depicted in Figure 6. To intuitively present the accuracy of the visual localization, the position and orientation results obtained by visual localization are compared with the results of the motion capture system Qualisys Track Manager (QTM). The desired path is $y = -0.1$, the start point of the snake robot is (1.0 m, 0.6 m), and the initial orientation is $\frac{\pi}{2}$.

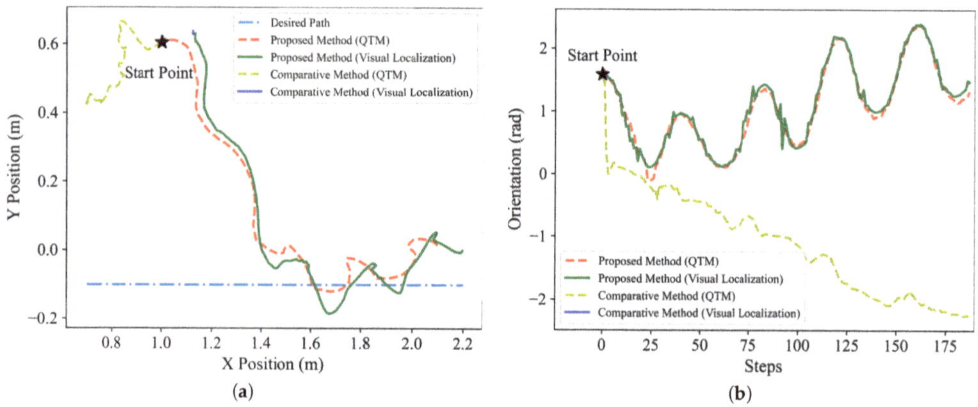

Figure 6. Results of visual localization for the trained policies with and without the visual stabilization item p_h. (**a**) Position results. (**b**) Orientation results.

From the results depicted in Figure 6, it is clear that the results of the visual localization are close to the actual results regardless of the position or orientation under the control of the proposed method. In contrast, under the control of the comparative method without the stabilization item p_h, the position and orientation of the snake robot (denoted by the solid blue lines shown in Figure 6) are only obtained at the beginning of the experiment, and subsequently, this information cannot be updated and the robot fails to follow the desired path, which is caused by the loss of visual target by the camera mounted on the head. At the beginning of the path following, the comparative method tends to generate a relatively large joint angle of the head to rapidly change the motion direction of the robot, thereby improving the convergence speed of the robot. However, the large head angle leads to a violent head swing, so the camera mounted on the head loses the visual target and thus, the pan-tilt cannot successfully compensate the joint angle of the head.

Subsequently, different actual experiments for a straight line and a circle are carried out to evaluate the following accuracy of the proposed hierarchical path following method. For the straight-line path following, $y = 0.0$ is taken as the desired path, and the start point of the snake robot is $(0.0\,\text{m}, -0.9\,\text{m})$. Then, circle $x^2 + y^2 = 4$ is chosen as the desired path, and the start point is $(-0.25\,\text{m}, 2.0\,\text{m})$. The following results are illustrated in Figures 7 and 8, which denote that the learned hierarchical path following policy can be directly transferred to the practical system and successfully drive the snake robot to converge to and then follow the desired paths.

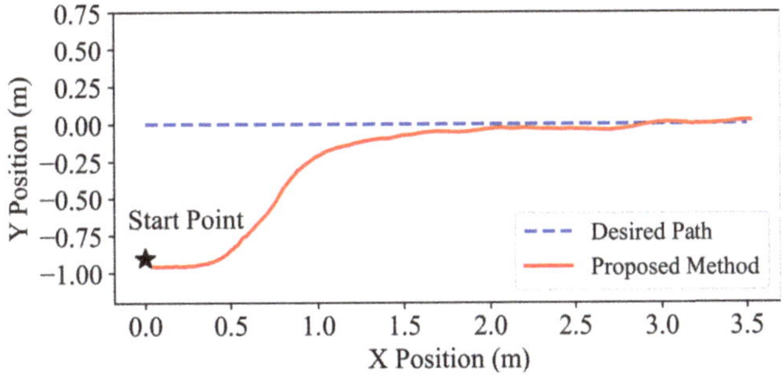

Figure 7. Path following results of the desired line for the proposed hierarchical method.

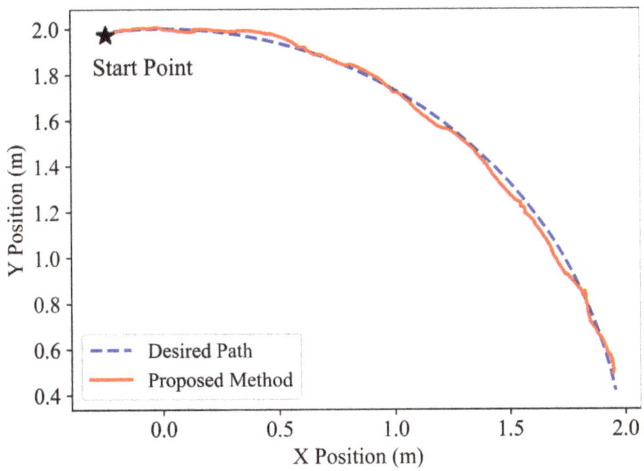

Figure 8. Path following results of the desired circle for the proposed hierarchical method.

4. Discussion

We can observe that the experimental results shown in Figure 8 exhibit the following performance of a snake robot for a quarter circle rather than an entire circle. This is because the number of the visual marker and the maximum rotation angle of the pan-tilt in our experiment are both limited; that is, there is only one fixed visual marker, and the rotation angle of the pan-tilt is limited to $(-140^\circ, 140^\circ)$. When the snake robot tracks the remaining three-quarters of the circle, the direction of the robot head will be opposite to the direction of the visual marker so that the angle that the pan-tilt needs to compensate for is larger than the maximum rotation angle, and the visual localization cannot be successfully completed.

Additionally, it is noteworthy that the accuracy of the path following in simulations and experiments mainly depends on the accuracy of localization. However, in this paper, an external visual marker is employed to assist positioning, which introduces additional systematic errors. In addition, the lack of diversity of the input data sources, i.e., only a monocular camera, limits the robustness and stability of the localization.

Future work: In future, we will focus on more intelligent methods to solve the problems mentioned above. Firstly, we will adopt multiple visual markers to assist the visual localization. When the direction of the head changes, a snake robot will autonomously select a visual marker in the corresponding direction to update its position so as to achieve the all-around visual positioning and track the path in any direction. In addition, we will focus on multi-sensor fusion technologies, where the sensors include GPS, IMU, camera, radar, etc., to complete autonomous perception without any external assistance, which will further improve the intelligence of perception and expand the application scenarios of snake robots. Finally, in order to further improve the robustness of the proposed method, some extrinsic perturbations will be imposed on a snake robot during training, including external force disturbance, sensory information noise and changes in physical parameters, thereby promoting the adaptability of a snake robot to system errors and changes in different application scenarios.

5. Conclusions

In this paper, a hierarchical RL-based control method is proposed to achieve satisfactory path following performance for snake robots with an onboard camera. Specifically, firstly, the hierarchical path following method, consisting of the RL policy training layer and the gait executive layer, optimizes the learning efficiency, exhibits reliable path following ability, and guarantees the transferability of the learned policy to the practical system by sufficiently combining the advantage of the RL policy network and the gait equation. Subsequently, the position of a snake robot can be updated in real time via visual localization due to the introduction of the visual localization stabilization item in the reward function. A series of simulation and hardware experimental results validate that the proposed method is capable of achieving a precise and fast convergence with respect to the path following tasks for a snake robot with autonomous visual perception.

Author Contributions: Conceptualization, X.G., Y.F. and L.L.; methodology, X.G., Y.F. and L.L.; software, L.L.; validation, L.L.; formal analysis, L.L.; investigation, L.L.; resources, L.L.; data curation, L.L.; writing—original draft preparation, L.L.; writing—review and editing, X.G. and Y.F.; visualization, L.L.; supervision, X.G. and Y.F.; project administration, X.G. and Y.F.; supervised the project, X.G. and Y.F.; funding acquisition, X.G. and Y.F. All authors have read and agreed to the published version of the manuscript.

Funding: This research was supported by the National Natural Science Foundation of China (62073176).

Institutional Review Board Statement: Not applicable.

Informed Consent Statement: Not applicable.

Data Availability Statement: Not applicable.

Conflicts of Interest: The authors declare no conflict of interest.

References

1. Hirose, S. *Biologically Inspired Robots: Snake-like Locomotors and Manipulators*; Oxford University Press: Oxford, UK, 1993.
2. Ye, C.; Ma, S.; Li, B.; Wang, Y. Turning and side motion of snake-like robot. In Proceedings of the IEEE International Conference on Robotics and Automation, 2004. Proceedings. ICRA '04. 2004, New Orleans, LA, USA, 26 April 2004–1 May 2004; Volume 5, pp. 5075–5080. [CrossRef]
3. Liljebäck, P.; Pettersen, K.Y.; Stavdahl, Ø.; Gravdahl, J.T. A review on modelling, implementation, and control of snake robots. *Robot. Auton. Syst.* **2012**, *60*, 29–40. [CrossRef]
4. Liljeback, P.; Haugstuen, I.U.; Pettersen, K.Y. Path following control of planar snake robots using a cascaded approach. *IEEE Trans. Control Syst. Technol.* **2011**, *20*, 111–126. [CrossRef]
5. Rezapour, E.; Pettersen, K.Y.; Liljebäck, P.; Gravdahl, J.T. Path following control of planar snake robots using virtual holonomic constraints. In Proceedings of the 2013 IEEE International Conference on Robotics and Biomimetics (ROBIO), Shenzhen, China, 12–14 December 2013; pp. 530–537. [CrossRef]
6. Wang, G.; Yang, W.; Shen, Y.; Shao, H. Adaptive path following of snake robot on ground with unknown and varied friction coefficients. In Proceedings of the 2018 IEEE/RSJ International Conference on Intelligent Robots and Systems (IROS), Madrid, Spain, 1–5 October 2018; pp. 7583–7588.
7. Yang, W.; Wang, G.; Shao, H.; Shen, Y. Spline based curve path following of underactuated snake robots. In Proceedings of the 2019 International Conference on Robotics and Automation (ICRA), Montreal, QC, Canada, 20–24 May 2019; pp. 5352–5358.
8. Cao, Z.; Zhang, D.; Zhou, M. Direction Control and Adaptive Path Following of 3-d Snake-Like Robot Motion. *IEEE Trans. Cybern.* **2021**, *52*, 10980–10987. [CrossRef] [PubMed]
9. Li, D.; Pan, Z.; Deng, H.; Hu, L. Adaptive path following controller of a multijoint snake robot based on the improved serpenoid curve. *IEEE Trans. Ind. Electron.* **2021**, *69*, 3831–3842. [CrossRef]
10. Mohammadi, A.; Rezapour, E.; Maggiore, M.; Pettersen, K.Y. Direction following control of planar snake robots using virtual holonomic constraints. In Proceedings of the 53rd IEEE Conference on Decision and Control, Los Angeles, CA, USA, 15–17 December 2014; pp. 3801–3808.

11. Rezapour, E.; Hofmann, A.; Pettersen, K.Y.; Mohammadi, A.; Maggiore, M. Virtual holonomic constraint based direction following control of planar snake robots described by a simplified model. In Proceedings of the 2014 IEEE Conference on Control Applications (CCA), Juan Les Antibes, France, 8–10 October 2014; pp. 1064–1071.
12. Rezapour, E.; Hofmann, A.; Pettersen, K.Y. Maneuvering control of planar snake robots based on a simplified model. In Proceedings of the 2014 IEEE International Conference on Robotics and Biomimetics (ROBIO 2014), Bali, Indonesia, 5–10 December 2014; pp. 548–555.
13. Mohammadi, A.; Rezapour, E.; Maggiore, M.; Pettersen, K.Y. Maneuvering control of planar snake robots using virtual holonomic constraints. *IEEE Trans. Control Syst. Technol.* **2015**, *24*, 884–899. [CrossRef]
14. Duan, Y.; Chen, X.; Houthooft, R.; Schulman, J.; Abbeel, P. Benchmarking deep reinforcement learning for continuous control. In Proceedings of the International Conference on Machine Learning, PMLR, New York, NY, USA, 20–22 June 2016; pp. 1329–1338.
15. Kiumarsi, B.; Vamvoudakis, K.G.; Modares, H.; Lewis, F.L. Optimal and autonomous control using reinforcement learning: A survey. *IEEE Trans. Neural Networks Learn. Syst.* **2017**, *29*, 2042–2062. [CrossRef] [PubMed]
16. Ibarz, J.; Tan, J.; Finn, C.; Kalakrishnan, M.; Pastor, P.; Levine, S. How to train your robot with deep reinforcement learning: Lessons we have learned. *Int. J. Robot. Res.* **2021**, *40*, 698–721. [CrossRef]
17. Iscen, A.; Caluwaerts, K.; Tan, J.; Zhang, T.; Coumans, E.; Sindhwani, V.; Vanhoucke, V. Policies modulating trajectory generators. In Proceedings of the Conference on Robot Learning, PMLR, Zürich, Switzerland, 29–31 October 2018; pp. 916–926.
18. Lee, J.; Hwangbo, J.; Wellhausen, L.; Koltun, V.; Hutter, M. Learning quadrupedal locomotion over challenging terrain. *Sci. Robot.* **2020**, *5*, eabc5986. [CrossRef] [PubMed]
19. Yang, Y.; Caluwaerts, K.; Iscen, A.; Zhang, T.; Tan, J.; Sindhwani, V. Data efficient reinforcement learning for legged robots. In Proceedings of the Conference on Robot Learning, PMLR, Cambridge, MA, USA, 16–18 November 2020; pp. 1–10.
20. Xie, Z.; Berseth, G.; Clary, P.; Hurst, J.; van de Panne, M. Feedback control for cassie with deep reinforcement learning. In Proceedings of the 2018 IEEE/RSJ International Conference on Intelligent Robots and Systems (IROS), Madrid, Spain, 1–5 October 2018; pp. 1241–1246.
21. Li, Z.; Cheng, X.; Peng, X.B.; Abbeel, P.; Levine, S.; Berseth, G.; Sreenath, K. Reinforcement learning for robust parameterized locomotion control of bipedal robots. In Proceedings of the 2021 IEEE International Conference on Robotics and Automation (ICRA), Xi'an, China, 30 May 2021–5 June 2021; pp. 2811–2817.
22. Liu, L.; Guo, X.; Fang, Y. Goal-driven Motion Control of Snake Robots with Onboard Cameras via Policy Improvement with Path Integrals. In Proceedings of the 2021 IEEE International Conference on Robotics and Biomimetics (ROBIO), Sanya, China, 27–31 December 2021; pp. 61–68. [CrossRef]
23. Ponte, H.; Queenan, M.; Gong, C.; Mertz, C.; Travers, M.; Enner, F.; Hebert, M.; Choset, H. Visual sensing for developing autonomous behavior in snake robots. In Proceedings of the 2014 IEEE International Conference on Robotics and Automation (ICRA), Hong Kong, China, 31 May 2014–7 June 2014, pp. 2779–2784.
24. Chang, A.H.; Feng, S.; Zhao, Y.; Smith, J.S.; Vela, P.A. Autonomous, monocular, vision-based snake robot navigation and traversal of cluttered environments using rectilinear gait motion. *arXiv* **2019**, arXiv:1908.07101.
25. Liu, L.; Xi, W.; Guo, X.; Fang, Y. Vision-based Path Following of Snake-like Robots. In Proceedings of the 2021 IEEE International Conference on Robotics and Automation (ICRA), Xi'an, China, 30 May 2021–5 June 2021; pp. 3084–3090. [CrossRef]
26. Todorov, E.; Erez, T.; Tassa, Y. MuJoCo: A physics engine for model-based control. In Proceedings of the 2012 IEEE/RSJ International Conference on Intelligent Robots and Systems, Vilamoura-Algarve, Portugal, 7–12 October 2012; pp. 5026–5033. [CrossRef]

sensors

MDPI

Article

The Effective Coverage of Homogeneous Teams with Radial Attenuation Models

Yuan-Rui Yang [1,*,†], Qiyu Kang [2,†] and Rui She [2,†]

[1] School of Mechanical and Aerospace Engineering, Nanyang Technological University, Singapore 637460, Singapore
[2] Centre for Information Sciences and Systems, Nanyang Technological University, Singapore 639798, Singapore
* Correspondence: yang_yuanrui@u.nus.edu
† Current address: Continental-NTU Corporate Lab, Nanyang Technological University, Singapore 637553, Singapore.

Abstract: For the area coverage (e.g., using a WSN), despite the comprehensive research works on full-plane coverage using a multi-node team equipped with the ideal constant model, only very few works have discussed the coverage of practical models with varying intensity. This paper analyzes the properties of the effective coverage of multi-node teams consisting of a given numbers of nodes. Each node is equipped with a radial attenuation disk model as its individual model of coverage, which conforms to the natural characteristics of devices in the real world. Based on our previous analysis of 2-node teams, the properties of the effective coverage of 3-node and n-node ($n \geq 4$) teams in regular geometric formations are analyzed as generalized cases. Numerical analysis and simulations for 3-node and n-node teams ($n \geq 4$) are conducted separately. For the 3-node cases, the relations between the side lengths of equilateral triangle formation and the effective coverage of the team equipped with two different types of models are respectively inspected. For the n-node cases ($n \geq 4$), the effective coverage of a team in three formations, namely regular polygon, regular star, and equilateral triangular tessellation (for $n = 6$), are investigated. The results can be applied to many scenarios, either dynamic (e.g., robots with sensors) or static, where a team of multiple nodes cooperate to produce a larger effective coverage.

Keywords: coverage control; multi-robot; cooperative sensors; optimization

Citation: Yang, Y.-R.; Kang, Q.; She, R. The Effective Coverage of Homogeneous Teams with Radial Attenuation Models. *Sensors* **2023**, *23*, 350. https://doi.org/10.3390/s23010350

Academic Editor: Felipe Jiménez

Received: 20 October 2022
Revised: 20 December 2022
Accepted: 23 December 2022
Published: 29 December 2022

1. Introduction

1.1. Background

Many static devices and autonomous robots (collectively named "nodes") can collaboratively sense or influence their surroundings. Coverage (area coverage) and effective coverage [1] are important metrics that reflect the sensing or influencing ability, or the performance of a node or a node team, such as a light array, air-conditioner team, wireless sensor network, etc. The analysis of a team's effective coverage depends on the problems to be solved and their adopted models of coverage. The most common categories of problems—optimal tessellation patterns (e.g., sensor placement issue) in a given plane have been extensively studied. Most of them have adopted the simplest binary disk model (having a constant value within the disk region centered at the node), which focuses on k-coverage [2,3]; only a very few have analyzed the effective coverage of cooperative sensors within an infinite plane using a non-binary (non-constant) sensing model [4,5].

Compared to this category of problems, the optimization of the effective coverage of a node team containing a given number of nodes in a certain operational setting (i.e., the inverse problem) has been rarely studied due to its difficulties in mathematics. For practical models with non-constant intensities (i.e., non-constant models), both maximizing the team's effective coverage and maintaining its connectedness are much more complicated than the ideal binary model [6,7] with which the nodes may move freely to independently produce some effective coverage. If the coverage intensity of one node is already larger than the requirement, the team can achieve its maximum coverage by simply avoiding any overlap among the individual coverage disks. However, for non-constant models, the area and the connectedness of a team's effective coverage have complicated relationships with the team's formation; any change in the team's formation causes changes in its effective coverage. However, it is useful in and can be applied to many dynamic problems involving moving robots, such as mobile WSNs, cooperative illumination, animal husbandry, and agriculture (for birds detection/driving), etc. [1,8]. Hence, maximizing the effective coverage of a node team and maintaining the connectedness of its effectively covered region are desirable in such practical applications, which motivates our study in this article.

1.2. Literature Review

The optimal tessellation pattern problem (to fully cover a plane) using the binary disk model (a constant non-zero intensity within the coverage disk) has been extensively studied (e.g., [2,3,9–14]) based on the binary disk coverage model. Many other works involving multi-node cooperative coverage (e.g., [15–17]) also adopt this model. However, such a model is too simple to characterize most realistic devices.

Some new non-constant models (with varying intensity) have been proposed [4,5,7,18] for sensing activities, where multiple sensors are cooperatively used to enlarge the sensing area and to improve the sensing accuracy. For example, in [5], Zhu et al. proposed that a point is "confident information covered" if the root-mean-square error of its estimated signal value is below the application requirement. Using attenuation sensing models, cooperative sensors can produce a larger "confident information coverage" than the sum of individual sensor coverages. These works found the optimal placement pattern (with the lowest sensor density) for an infinite plane under their model assumption. This model was also adopted in several later works [19,20]. Yáng et al. proposed the concept of effective coverage based on the coverage intensity to characterize the coverage behavior of practical devices equipped with attenuation models [1]. Their model follows the superposition principle, which focuses on influencing activities (output coverage) and can approximate the behavior of sensing activities. Effective coverages of 2-node teams equipped with different models were analyzed. Meng et al. proposed a learning method for the effective coverage of points of interest, where the sensing quality of an agent is a decreasing function of the distance from this agent [21].

In this paper, we study the effective coverage of n-node homogeneous teams with $n \geq 3$. The nodes are equipped with radial attenuation disk models to approximate realistic devices. A team's (cooperative) coverage follows the superposition principle. Optimal separation distances that maximize a team's effective coverage are analyzed and simulated, respectively, for two types of models of coverage. The results can be directly applied to practical applications such as mobile air-conditioner teams [22] and birds driving robots, for output applications [1]; and for input applications, e.g., mobile wireless sensor networks, the approximation using our models and analysis is much more accurate and practical than the existing binary models. To the best of our knowledge, this is the first work to study the effective coverage optimization of a team containing a given number ($\geq$3) of nodes equipped with a non-constant model of coverage.

The rest of the paper is organized as follows. Section 2 formulates our model and then analyzes 3-node teams and n-node teams, respectively. The analysis and simulation of some example models are presented in Section 3. Finally, we conclude in Section 4.

2. Coverage of an n-Node Team in Regular Polygon Formation

The situation of an n-node team with $n \geq 3$ is much more complicated compared to 2-node teams. It is hard to find their explicit analytical relationship (function) with the team's effective coverage, and there are infinite possibilities of formations on a 2D plane.

Therefore, for $n > 3$, only some symmetric cases are analyzed, where the teams are in regular polygon and regular star formations.

We consider a set of n nodes located at planar positions s_i with $i = 1, 2, \ldots, n$. As the vertices of a polygon, they are in the formation of an n-sided regular polygon with a circumradius R. This set of nodes can be represented as a network $S = \{s_1, \ldots, s_n\}$. We establish a polar coordinate system so that the pole is at the circumcenter O of the n-sided regular polygon, and the polar axis crosses the node s_n. We use D to represent the side length of the polygon formation, i.e., the separation of any two adjacent nodes on the polygon vertices. So, $s_i = \left(R, \frac{2\pi i}{n}\right)$ and $R = 0.5D\csc(\pi/n)$, as illustrated in Figure 1. The distance between any point $p(\rho \geq 0, \theta)$ and node s_i located at the polygon vertex $\left(R, \frac{2\pi i}{n}\right)$ is then $d_i = \sqrt{R^2 + \rho^2 - 2R\rho\cos\left(\theta - \frac{2\pi i}{n}\right)}$.

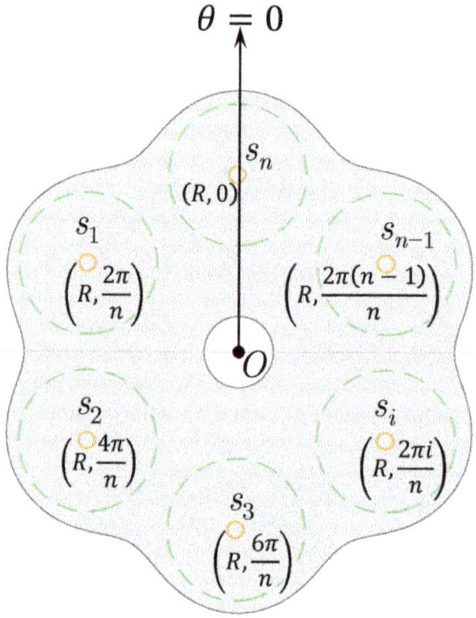

Figure 1. Illustration of an n-node team forming an n-sided regular polygon, with the nodes being located on its vertices. The shaded gray region is the effectively covered region (an example) by the team, while the n dashed circles show individual effective coverages with radius $\bar{r}$. The polar coordinate system is established in such a way where the pole is located at the circumcenter of the polygon, and the polar axis points towards s_n.

The coverage intensity induced by a node $s_i \in S$ at a point $p \in \mathbb{R}^2$ is represented as a non-constant function $I(d_i) : \mathbb{R} \to \mathbb{R}$, where $d_i = \|p - s_i\|$ is the distance from s_i to p. The coverage of a node s_i (denoted by C_i) is $C_i = \iint_{\mathbb{R}^2} \delta_i(p)\,\mathrm{d}p$, where $\delta_i(p) = 1$ if $I(d_i) > 0$ and $\delta_i(p) = 0$ otherwise. The individual coverage radius is denoted by $r \in \mathbb{R}^+$. With a certain threshold $\gamma \in \mathbb{R}^+$, the effective coverage of a node s_i (denoted by $\bar{C}_i$) is $\bar{C}_i = \iint_{\mathbb{R}^2} \bar{\delta}_i(p)\,\mathrm{d}p$, where $\bar{\delta}_i(p) = 1$ if $I(d_i) \geq \gamma$, and $\bar{\delta}_i(p) = 0$ otherwise. From this definition, we know that $\bar{C}_S = \iint_{\mathbb{R}^2} \bar{\delta}(p)\,\mathrm{d}p$, where $\bar{\delta}(p) = 1$ if $\Phi(p) \geq \gamma$ and $\bar{\delta}(p) = 0$ otherwise. For one team, there are two critical separation distances:

1. "Max distance", which yields the maximum team's effective coverage.
2. "Last-connection distance". Any separation D greater than this distance will make the team's effectively covered region disconnect into more than one part.

2.1. Three-Node Teams

For a three-node team forming an equilateral triangle, the nodes are located at the polar points $(R, 2\pi/3)$, $(R, -2\pi/3)$, $(R, 0)$. The coverage intensity at any point $p(\rho \geq 0, \theta)$ induced by the team is

$$\begin{aligned}\Phi(p) =& I\left(\sqrt{R^2+\rho^2-2R\rho\cos\left(\theta-\frac{2}{3}\pi\right)}\right)\\ &+I\left(\sqrt{R^2+\rho^2-2R\rho\cos\left(\theta+\frac{2}{3}\pi\right)}\right)+I\left(\sqrt{R^2+\rho^2-2R\rho\cos\theta}\right)\end{aligned} \tag{1}$$

and its first-order and second-order partial derivative w.r.t. ρ are

$$\begin{aligned}\frac{\partial\Phi(p)}{\partial\rho} =& \left.\frac{\mathrm{d}I}{\mathrm{d}d}\right|_{d=\sqrt{R^2+\rho^2-2R\rho\cos(\theta-\frac{2}{3}\pi)}}\frac{\rho-R\cos(\theta-\frac{2}{3}\pi)}{\sqrt{R^2+\rho^2-2R\rho\cos(\theta-\frac{2}{3}\pi)}}\\ &+\left.\frac{\mathrm{d}I}{\mathrm{d}d}\right|_{d=\sqrt{R^2+\rho^2-2R\rho\cos(\theta+\frac{2}{3}\pi)}}\frac{\rho-R\cos(\theta+\frac{2}{3}\pi)}{\sqrt{R^2+\rho^2-2R\rho\cos(\theta+\frac{2}{3}\pi)}}\\ &+\left.\frac{\mathrm{d}I}{\mathrm{d}d}\right|_{d=\sqrt{R^2+\rho^2-2R\rho\cos\theta}}\frac{\rho-R\cos\theta}{\sqrt{R^2+\rho^2-2R\rho\cos\theta}},\end{aligned} \tag{2}$$

$$\begin{aligned}\frac{\partial^2\Phi(p)}{\partial\rho^2} =&\\ &\left.\frac{\mathrm{d}^2I(d)}{\mathrm{d}d^2}\right|_{d=d_1}\frac{\left[\rho-R\cos(\theta-\frac{2\pi}{3})\right]^2}{R^2+\rho^2-2R\rho\cos(\theta-\frac{2\pi}{3})}+\left.\frac{\mathrm{d}I(d)}{\mathrm{d}d}\right|_{d=d_1}\frac{R^2\sin^2(\theta-\frac{2\pi}{3})}{\left[R^2+\rho^2-2R\rho\cos(\theta-\frac{2\pi}{3})\right]^{1.5}}\\ &+\left.\frac{\mathrm{d}^2I(d)}{\mathrm{d}d^2}\right|_{d=d_2}\frac{\left[\rho-R\cos(\theta+\frac{2\pi}{3})\right]^2}{R^2+\rho^2-2R\rho\cos(\theta+\frac{2\pi}{3})}+\left.\frac{\mathrm{d}I(d)}{\mathrm{d}d}\right|_{d=d_2}\frac{R^2\sin^2(\theta+\frac{2\pi}{3})}{\left[R^2+\rho^2-2R\rho\cos(\theta+\frac{2\pi}{3})\right]^{1.5}}\\ &+\left.\frac{\mathrm{d}^2I(d)}{\mathrm{d}d^2}\right|_{d=d_3}\frac{\left[\rho-R\cos\theta\right]^2}{R^2+\rho^2-2R\rho\cos\theta}+\left.\frac{\mathrm{d}I(d)}{\mathrm{d}d}\right|_{d=d_3}\frac{R^2\sin^2\theta}{\left[R^2+\rho^2-2R\rho\cos\theta\right]^{1.5}}\end{aligned} \tag{3}$$

where $d_1 = \sqrt{R^2+\rho^2-2R\rho\cos(\theta-\frac{2}{3}\pi)}$, $d_2 = \sqrt{R^2+\rho^2-2R\rho\cos(\theta+\frac{2}{3}\pi)}$, and $d_3 = \sqrt{R^2+\rho^2-2R\rho\cos\theta}$. At the circumcenter $O(0,0)$,

$$\Phi(O) = 3I(R) \tag{4}$$

$$\left.\frac{\partial\Phi(p)}{\partial\rho}\right|_{\rho=0} = -\left.\frac{\mathrm{d}I}{\mathrm{d}d}\right|_{d=R}\left[\cos\left(\theta-\frac{2}{3}\pi\right)+\cos\left(\theta+\frac{2}{3}\pi\right)+\cos\theta\right]=0 \tag{5}$$

Obviously, if $I(d)$ is a concave model, $\mathrm{d}I(d)/\mathrm{d}d < 0$ and $\mathrm{d}^2 I(d)/\mathrm{d}d^2 < 0$, and Equation (6) is negative; otherwise if $I(d)$ is a convex model, Equation (6) is positive; see Appendix B for the proof. Hence, the circumcenter O is a local minimum point of coverage intensity when the nodes are equipped with a convex model of coverage; and it is a local maximum point of coverage intensity when the nodes are equipped with a concave model of coverage and $R < r$, i.e., the coverage overlapping region of the n nodes exists.

Next, we establish the general approach for finding the "last-connection" distance and the coordinate of its corresponding connection point. We only discuss the situations where the nodes are equipped with the unbounded convex model as an example. For the truncated convex models and concave models, the methods are similar to the unbounded models in the overlapping region centered at $O(0,0)$, and the same with their corresponding 2-node cases [1] out of that overlapping region.

$$
\begin{aligned}
\left.\frac{\partial^2\Phi(p)}{\partial\rho^2}\right|_{\rho=0} =& \left.\frac{\mathrm{d}^2 I(d)}{\mathrm{d}d^2}\right|_{d=R}\cos^2\left(\theta-\frac{2}{3}\pi\right)+\left.\frac{\mathrm{d}I(d)}{\mathrm{d}d}\right|_{d=R}\frac{\sin^2(\theta-\frac{2}{3}\pi)}{R} \\
&+\left.\frac{\mathrm{d}^2 I(d)}{\mathrm{d}d^2}\right|_{d=R}\cos^2\left(\theta+\frac{2}{3}\pi\right)+\left.\frac{\mathrm{d}I(d)}{\mathrm{d}d}\right|_{d=R}\frac{\sin^2(\theta+\frac{2}{3}\pi)}{R} \\
&+\left.\frac{\mathrm{d}^2 I(d)}{\mathrm{d}d^2}\right|_{d=R}\cos^2\theta+\left.\frac{\mathrm{d}I(d)}{\mathrm{d}d}\right|_{d=R}\frac{\sin^2\theta}{R} \\
=& \left.\frac{\mathrm{d}^2 I(d)}{\mathrm{d}d^2}\right|_{d=R}\frac{3+\cos\left(2\theta-\frac{4}{3}\pi\right)+\cos\left(2\theta+\frac{4}{3}\pi\right)+\cos 2\theta}{2} \\
&+\left.\frac{\mathrm{d}I(d)}{\mathrm{d}d}\right|_{d=R}\frac{3-\cos\left(2\theta-\frac{4}{3}\pi\right)-\cos\left(2\theta+\frac{4}{3}\pi\right)-\cos 2\theta}{2R} \\
=& \left.\frac{\mathrm{d}^2 I(d)}{\mathrm{d}d^2}\right|_{d=R}\frac{3}{2}+\left.\frac{\mathrm{d}I(d)}{\mathrm{d}d}\right|_{d=R}\frac{3}{2R} \\
=& \frac{3}{2}\left[\left.\frac{\mathrm{d}^2 I(d)}{\mathrm{d}d^2}\right|_{d=R}+\left.\frac{\mathrm{d}I(d)}{\mathrm{d}d}\right|_{d=R}\frac{1}{R}\right]
\end{aligned} \tag{6}
$$

From the property of the convex decreasing functions, it is obvious that with any fixed θ value and $\rho \in [0, +\infty)$, $\Phi(p)$ has only two local extrema. The local minimum happens at $O(0,0)$, which we already derived above. When ρ increases, $\Phi(p)$ increases up to its global maximum value and then decreases asymptotically to zero (see Figure 2). Based on symmetry, the "last-connection" points must be on the mirror line between every two nodes: their angle coordinates $\theta = \pi/3, \pi, -\pi/3$, as shown in Figure 3. So, the ρ coordinate of these "last-connection" points can be found by determining the global maximum point of $\Phi(p)$ on any of the three rays $\theta = \pi/3, \pi, -\pi/3$. Let's look at the "last-connection" point $q_{1,2}(\rho, \pi)$ on the ray $\theta = \pi$ (between s_1 and s_2), where the coverage intensity and its first-order derivative w.r.t. ρ are:

$$
\Phi(q_{1,2}) = 2I\left(\sqrt{R^2+\rho^2-R\rho}\right) + I(R+\rho) \tag{7}
$$

$$
\frac{\partial\Phi(q_{1,2})}{\partial\rho} = \left.\frac{\mathrm{d}I}{\mathrm{d}d}\right|_{d=\sqrt{R^2+\rho^2-R\rho}}\left(\frac{2\rho-R}{\sqrt{R^2+\rho^2-R\rho}}\right)+\left.\frac{\mathrm{d}I}{\mathrm{d}d}\right|_{d=R+\rho} \tag{8}
$$

The "last-connection distance" ($D = \sqrt{3}R$) and the ρ coordinate of its corresponding "last-connection" point can be found through solving the following equations:

$$
\begin{cases}
\Phi(q_{1,2}) = \gamma \\
\dfrac{\partial\Phi(q_{1,2})}{\partial\rho} = 0 \\
\rho > 0
\end{cases} \tag{9}
$$

An example analysis will be provided in Section 3.

2.2. General n-Node Team

After analyzing the effective coverage of 3-node teams, we now generalize it to n-node teams in n-sided polygon formations. We apply the usual notion of topological simple connectedness to characterize the effective coverage of a team. Consider the effectively covered region of team S. If there exist one or more "holes" within this region that are not effectively covered by S, then we say that the effectively covered region of S is not simply connected. Next, we establish a theory that enables us to easily examine whether the effectively covered region of the team equipped with an unbounded convex model of coverage is simply connected, i.e., whether any hole (that is not effectively covered) exists as a subregion within the effectively covered region.

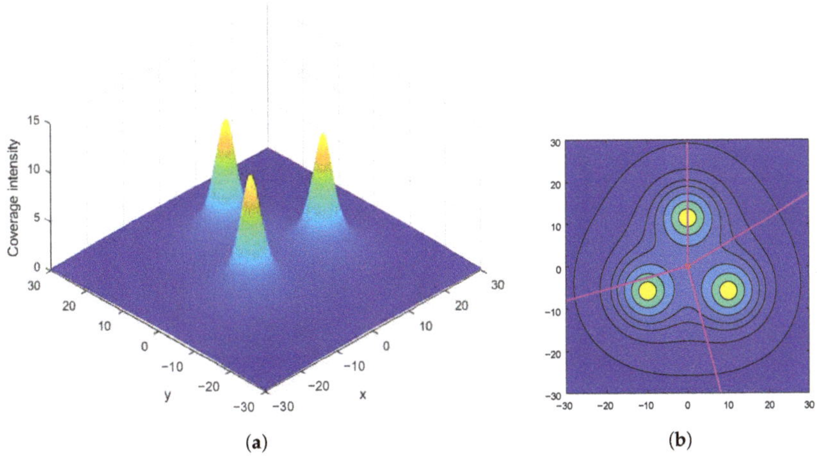

Figure 2. Coverage intensity of a 3-node team forming an equilateral triangle, with the nodes being located on the vertices. (**a**) Three-dimensional illustration, where the z-axis is the coverage intensity and the three peaks locate the three nodes; (**b**) contour plot with four arbitrary example rays.

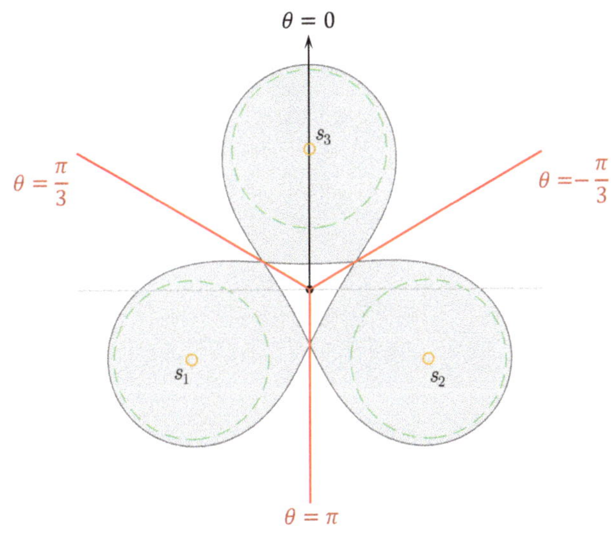

Figure 3. The "last-connection" points are on the mid-perpendicular line of every two adjacent nodes.

Theorem 1. *Weakest Point Theorem.*
For an n-node team $S = \{s_1, s_2, \ldots, s_n\}$ in the formation of an n-sided regular polygon where each node is equipped with a convex model of coverage and $n \geq 3$, its circumcenter O is effectively covered if and only if the entire effectively covered region is simply connected.

Proof. The sufficient condition is obvious. We only prove the necessary condition, which is equivalent to proving that the circumcenter $O(0,0)$ is the only local minimum of coverage intensity Φ along any radial direction of the polar coordinate system (i.e., for any fixed θ value).

The first-order partial derivatives of I w.r.t. ρ and θ are

$$\begin{cases} \dfrac{\partial I(d_i)}{\partial \rho} = \dfrac{\mathrm{d}I(d_i)}{\mathrm{d}d}\dfrac{\partial d}{\partial \rho} = \dfrac{\mathrm{d}I(d_i)}{\mathrm{d}d}\dfrac{\rho - R\cos\left(\theta - \frac{2\pi i}{n}\right)}{\sqrt{R^2+\rho^2-2R\rho\cos\left(\theta-\frac{2\pi i}{n}\right)}} \\ \dfrac{\partial I(d_i)}{\partial \theta} = \dfrac{\mathrm{d}I(d_i)}{\mathrm{d}d}\dfrac{\partial d}{\partial \theta} = \dfrac{\mathrm{d}I(d_i)}{\mathrm{d}d}\dfrac{R\rho\sin\left(\theta - \frac{2\pi i}{n}\right)}{\sqrt{R^2+\rho^2-2R\rho\cos\left(\theta-\frac{2\pi i}{n}\right)}} \end{cases} \tag{10}$$

When $\rho = 0$, i.e., at the circumcenter $O(0,0)$,

$$\left.\frac{\partial \Phi(p)}{\partial \rho}\right|_{\rho=0} = \sum_{i=1}^{n}\frac{\partial I(d_i)}{\partial \rho} = -\left.\frac{\mathrm{d}I(d)}{\mathrm{d}d}\right|_{d=R}\sum_{i=1}^{n}\cos\left(\theta - \frac{2\pi i}{n}\right) = 0 \tag{11}$$

for any θ. See Appendix A for proof.

Then, the second-order partial derivative w.r.t. ρ is

$$\begin{aligned} \frac{\partial^2 I(d_i)}{\partial \rho^2} &= \frac{\frac{\mathrm{d}I(d_i)}{\mathrm{d}d}}{\partial \rho}\frac{\partial d}{\partial \rho} + \frac{\mathrm{d}I(d_i)}{\mathrm{d}d}\frac{\partial^2 d}{\partial \rho^2} \\ &= \frac{\mathrm{d}^2 I(d_i)}{\mathrm{d}d^2}\frac{\left[\rho - R\cos\left(\theta-\frac{2\pi i}{n}\right)\right]^2}{R^2+\rho^2-2R\rho\cos\left(\theta-\frac{2\pi i}{n}\right)} + \frac{\mathrm{d}I(d_i)}{\mathrm{d}d}\frac{R^2\sin^2\left(\theta-\frac{2\pi i}{n}\right)}{\left[R^2+\rho^2-2R\rho\cos\left(\theta-\frac{2\pi i}{n}\right)\right]^{1.5}}. \end{aligned} \tag{12}$$

So, at the circumcenter O,

$$\begin{aligned} \left.\frac{\partial^2 \Phi(p)}{\partial \rho^2}\right|_{\rho=0} &= \left.\frac{\mathrm{d}^2 I(d)}{\mathrm{d}d^2}\right|_{d=R}\sum_{i=1}^{n}\cos^2\left(\theta-\frac{2\pi i}{n}\right) + \left.\frac{\mathrm{d}I(d)}{\mathrm{d}d}\right|_{d=R}\frac{1}{R}\sum_{i=1}^{n}\sin^2\left(\theta-\frac{2\pi i}{n}\right) \\ &= \left.\frac{\mathrm{d}^2 I(d)}{\mathrm{d}d^2}\right|_{d=R}\sum_{i=1}^{n}\frac{1+\cos\left(2\theta-\frac{4\pi i}{n}\right)}{2} + \left.\frac{\mathrm{d}I(d)}{\mathrm{d}d}\right|_{d=R}\frac{1}{R}\sum_{i=1}^{n}\frac{1-\cos\left(2\theta-\frac{4\pi i}{n}\right)}{2} \\ &= \left.\frac{\mathrm{d}^2 I(d)}{\mathrm{d}d^2}\right|_{d=R}\frac{n}{2} + \left.\frac{\mathrm{d}I(d)}{\mathrm{d}d}\right|_{d=R}\frac{n}{2R} \\ &= \frac{n}{2}\left[\left.\frac{\mathrm{d}^2 I(d)}{\mathrm{d}d^2}\right|_{d=R} + \frac{1}{R}\left.\frac{\mathrm{d}I(d)}{\mathrm{d}d}\right|_{d=R}\right] \\ &> 0 \end{aligned} \tag{13}$$

Therefore, the circumcenter O is the local minimum point of the coverage intensity.

On the other hand obviously, when $I(d)$ is concave and $d_i < r$, Equation (12) is always negative. Therefore,

$$\frac{\partial^2 \Phi(p)}{\partial \rho^2} = \sum_{i=1}^{n}\frac{\partial^2 I(d_i)}{\partial \rho^2} < 0 \text{ when } R < r \tag{14}$$

which reveals that the coverage intensity $\Phi(O)$ at the circumcenter is the local maximum value in the coverage overlapping region of the n nodes equipped with a concave model. □

For a team where each node is equipped with an unbounded convex model of coverage, Theorem 1 implies that the circumcenter of the polygon formation is the weakest point in terms of the intensity throughout the team's effectively covered region. When the separation distance D (side length of the polygon) increases, the effective coverage hole will appear at the circumcenter. The critical separation can be simply determined from $D = 2I^{-1}(\gamma/n)\sin(\pi/n)$. Besides determining the hole existence of the effectively covered region of S, with Theorem 1, we can also determine the "last-connection distance" and its corresponding "last-connection" points by solving Equation (9), for this n-node team.

Although the connectedness of the effectively covered region is clearly analyzed, it is still impossible to find the explicit relation between team formation and the maximum team's effective coverage. Some examples with explicit $I(d)$ expressions will be derived, simulated, and compared in the Simulations section below.

3. Simulations

For 3-node teams, unbounded convex models and concave models are discussed, and the teams are in equilateral triangle formations. For n-node teams where $n \geq 4$, both regular polygon and regular star formations are involved, and equilateral triangular tessellation is simulated for the 6-node team. Optimal separations that maximize the corresponding effective coverages are found. Example plots of intensity vs. separation are presented. All simulations are performed in Matlab R2022a on Linux (Debian 10) OS, Intel Core i9-12900KF CPU.

3.1. Unbounded Convex Model for a 3-Node Team in Equilateral Triangle Formation

Let's analyze the most common convex model

$$I(d) = \frac{\beta}{d^\lambda + \alpha} \tag{15}$$

where the scaling factor and the exponent $\beta, \lambda \in \mathbb{R}^+$, and the offset constant $\alpha \geq 0$. Its first-order derivative is

$$\frac{\mathrm{d}I}{\mathrm{d}d} = -\frac{\beta\lambda d^{\lambda-1}}{(d^\lambda + \alpha)^2} \tag{16}$$

and the individual effective coverage radius is $\bar{r} = \sqrt[\lambda]{\beta/\gamma - \alpha}$. Substitute them into Equation (9):

$$\begin{aligned}
\frac{\mathrm{d}\Phi(q_{1,2})}{\mathrm{d}\rho} &= \left.\frac{\mathrm{d}I}{\mathrm{d}d}\right|_{d=\sqrt{R^2+\rho^2-R\rho}} \left(\frac{2\rho - R}{\sqrt{R^2+\rho^2-R\rho}}\right) + \left.\frac{\mathrm{d}I}{\mathrm{d}d}\right|_{d=R+\rho} \\
&= -\frac{\beta\lambda(R^2+\rho^2-R\rho)^{\frac{\lambda-1}{2}}}{\left((R^2+\rho^2-R\rho)^{\frac{\lambda}{2}}+\alpha\right)^2} \frac{2\rho - R}{\sqrt{R^2+\rho^2-R\rho}} - \frac{\beta\lambda(R+\rho)^{\lambda-1}}{\left((R+\rho)^\lambda+\alpha\right)^2} \\
&= -\beta\lambda\left[\frac{(R^2+\rho^2-R\rho)^{\frac{\lambda}{2}-1}}{\left((R^2+\rho^2-R\rho)^{\frac{\lambda}{2}}+\alpha\right)^2}(2\rho - R) + \frac{(R+\rho)^{\lambda-1}}{\left((R+\rho)^\lambda+\alpha\right)^2}\right] = 0
\end{aligned} \tag{17}$$

Since $\beta \neq 0$ and $\lambda \neq 0$, we obtain

$$\begin{cases}
\dfrac{2}{(R^2+\rho^2-R\rho)^{\lambda/2}+\alpha} + \dfrac{1}{(R+\rho)^\lambda+\alpha} = \dfrac{1}{\bar{r}^\lambda+\alpha} \\
\dfrac{(R^2+\rho^2-R\rho)^{\frac{\lambda}{2}-1}}{\left((R^2+\rho^2-R\rho)^{\frac{\lambda}{2}}+\alpha\right)^2}(2\rho - R) + \dfrac{(R+\rho)^{\lambda-1}}{\left((R+\rho)^\lambda+\alpha\right)^2} = 0 \\
\rho > 0
\end{cases} \tag{18}$$

This is a set of hyper equations when $\lambda \neq 2$ or $\alpha \neq 0$. To make the equations solvable without losing universality, we set $\lambda = 2$ and $\alpha = 0$, and the equations above become

$$\begin{cases} \dfrac{2}{R^2+\rho^2-R\rho+0} + \dfrac{1}{R^2+\rho^2+2R\rho+0} = \dfrac{1}{\bar{r}^2+0} \\ \dfrac{2\rho-R}{(R^2+\rho^2-R\rho+0)^2} + \dfrac{R+\rho}{((R+\rho)^2+0)^2} = 0 \\ \rho > 0 \end{cases} \quad (19)$$

After some algebras, we obtain

$$\begin{cases} R^4+\rho R^3-3\bar{r}^2R^2+(\rho^2-3\bar{r}^2)\rho R+\rho^4-3\bar{r}^2\rho^2=0 \\ \rho(\rho^3+R\rho^2+2R^2\rho-R^3)=0 \\ \rho>0 \end{cases} \quad (20)$$

The two real roots of the second equation are $\rho = 0$ and $0.3926R$. Substitute $\rho = 0.3926R$ back to the first equation. Since $R \neq 0$, we get

$$1.4770R^2 - 4.6405\bar{r}^2 = 0 \quad (21)$$

Hence,

$$R = 1.7725\bar{r} \quad (22)$$

Therefore, for a 3-node team in equilateral triangle formation, the effectively covered region's "last-connection" distance between the nodes (i.e., the side length of this equilateral triangle) happens at $D = \sqrt{3}R = 3.0701\bar{r}$. At this critical separation distance, the whole region is only connected at three points $(0.6960\bar{r}, \pi/3)$, $(0.6960\bar{r}, \pi)$, $(0.6960\bar{r}, -\pi/3)$. Any distance farther than this will result in the team's effectively covered region being disconnected into three unconnected parts.

3.2. Concave Model for a 3-Node Team in Equilateral Triangle Formation

Many sensors follow concave models within their near field (short distances from the sensor). We use the quadratic difference equation to approximate the individual model of coverage:

$$I(d) = \begin{cases} \alpha(r^2-d^2) & , \ d \leq r \\ 0 & , \ d > r \end{cases} \quad (23)$$

where $\alpha \in \mathbb{R}^+$. The superposed coverage intensity in the overlapping region of the three nodes is

$$\Phi(p(x,y)) = \alpha\left(3r^2-3x^2-3y^2-D^2\right) \quad (24)$$

Let $\Phi(p(x,y)) \geq \gamma$; with individual effective coverage radius $\bar{r}^2 = r^2 - \gamma/\alpha$, we obtain the increased effective coverage region boundary in the overlapping region of the three nodes as

$$x^2+y^2 \leq \frac{2r^2+\bar{r}^2-D^2}{3} \quad (25)$$

when $R < r$, which is still a disk or part of a disk region.

Next, we simulate this model with different D and $\bar{r}$ values. The scaling factor $\alpha = 1$ and the individual coverage radius $r = 10$. $\bar{r} = 0.1r$ to $0.9r$ and $D = 2\bar{r}$ to $\bar{r} + r$ are examined, whose team's effective coverage is meshed in Figure 4. The optimal separation D_{m} vs. these $\bar{r}$ values are plotted in Figure 5. The effective coverages vs. D under three thresholds $\bar{r} = 0.3r, 0.5r, 0.7r$ are selected to be plotted in Figure 6, and their effectively covered regions are shown in Figure 7.

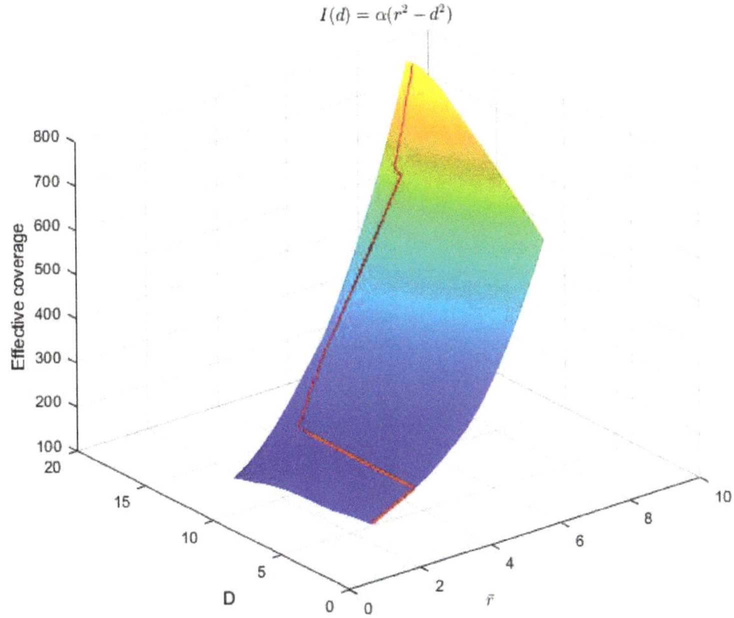

Figure 4. Three-dimensional mesh of team's effective coverage vs. separation D and $\bar{r}$. Nodes are equipped with concave models. $r = 10$. $\alpha = 1$. The red curve illustrates the maximum effective coverage at every $\bar{r}$ value.

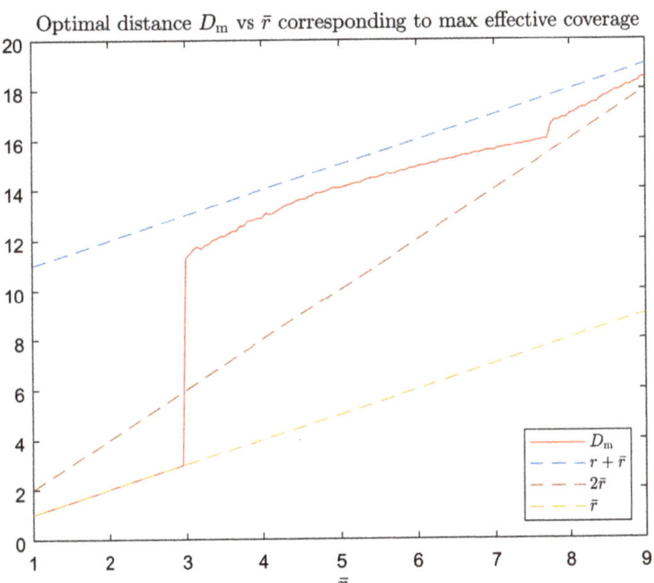

Figure 5. Relation between the optimal D and $\bar{r}$ corresponding to the maximum team's effective coverage.

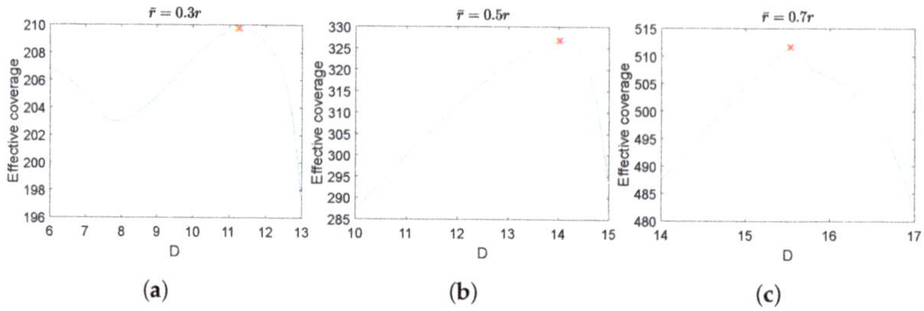

Figure 6. Team's effective coverage vs. separation D under three different thresholds: (**a**) $\bar{r} = 0.3r$, (**b**) $\bar{r} = 0.5r$, (**c**) $\bar{r} = 0.7r$. Nodes are equipped with concave models. $r = 10$. $\alpha = 1$. The red cross in each plot marks the maximum effective coverage.

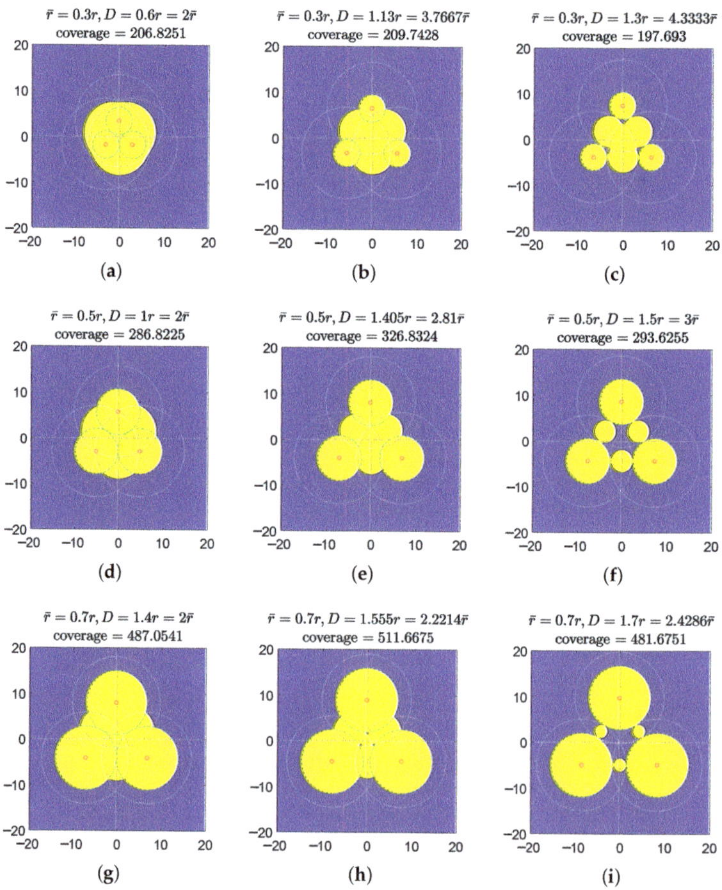

Figure 7. Illustration of resultant effective coverages of the 3-node team equipped with concave models $I(d) = \alpha(r^2 - d^2)$ under different thresholds γ (intuitively reflected as $\bar{r}$). Each row represents one value of γ: they are $\bar{r} = 0.3r, 0.5r, 0.7r$, respectively. Each column represents a different separation distance D: the first column (**a**,**d**,**g**) is the "tangent column", where the individual effectively covered regions in light green dashed circles are tangent to each other, i.e., $D = 2\bar{r}$; the second column (**b**,**e**,**h**) is the "max column", where the team's effective coverage is maximized with corresponding separations D; the third column (**c**,**f**,**i**) shows configurations where $D = r + \bar{r}$. $r = 10$. $\alpha = 1$.

As we can see from Figure 7, the maximum effective coverage may or may not have effective coverage hole(s). In the case that the threshold γ is low (resulting in large $\bar{r}$), it will have holes. When separation D increases, three effective coverage holes first appear at the three corners of the 3-node overlapping region, and then merge into one hole; after the effectively covered region in the center overlapping region disappears, the entire effectively covered region becomes simply the combination of the 2-node case.

From Figures 4 and 5, it is clearly seen that when the threshold γ is high ($\bar{r}$ is small), the optimal separation is $\bar{r}$; when $\bar{r}$ is roughly larger than $0.3r$, the optimal separation D_{m} falls between $2\bar{r}$ and $r + \bar{r}$. Figure 8 shows that the coverage intensity has seven local maxima located at the three nodes, three midpoints between every two nodes, and the circumcenter.

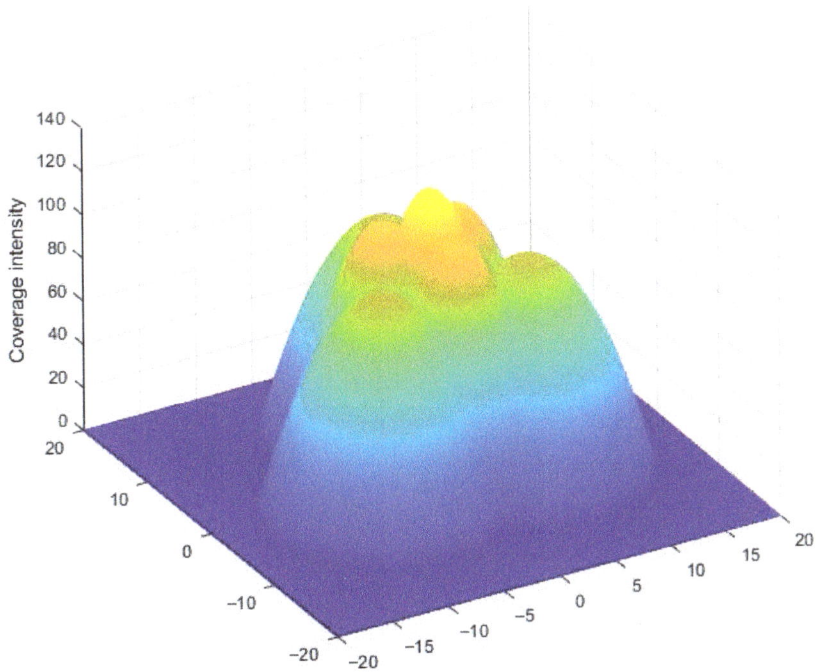

Figure 8. Three-dimensional illustration of coverage intensity of the 3-node team equipped with a concave model of coverage, in an equilateral triangle formation, with the nodes located on its vertices. The z-axis is the coverage intensity. $r = 10$. $\alpha = 1$.

3.3. *n* Nodes in Regular Polygon and Regular Star Formations

For n-node teams where $n \geq 4$, n-sided regular polygons and $(n-1)$-vertex regular star formations, and the 6-node team in equilateral triangular tessellation formation are simulated and compared. The term "$(n-1)$-vertex regular star" refers to the formation of an $(n-1)$-sided regular polygon with another node placed at its circumcenter. Each node is equipped with the unbounded convex model $I(d) = \beta/d^2$, where $\beta \in \mathbb{R}^+$. In this subsection, the scaling factor $\beta = 1$.

The effectively covered regions of 4,5,6,7-node teams in the polygon and star formations are illustrated in Figures 9 and 10, respectively. In both the regular polygon and regular star formations, the maximum teams' effective coverages are achieved when the individual effective coverage disks are apart from each other (neither tangent nor secant), but the entire effectively covered region is still connected. In regular polygon formations, an effective coverage hole is observed in each maximum effective coverage illustration at the circumcenter, which also verified in Theorem 1. Additionally, the effectively covered regions of the six-node team in equilateral triangle formation are illustrated in Figure 11.

Its maximum effective coverage is found to be slightly larger than the previous two formations, with $n = 6$.

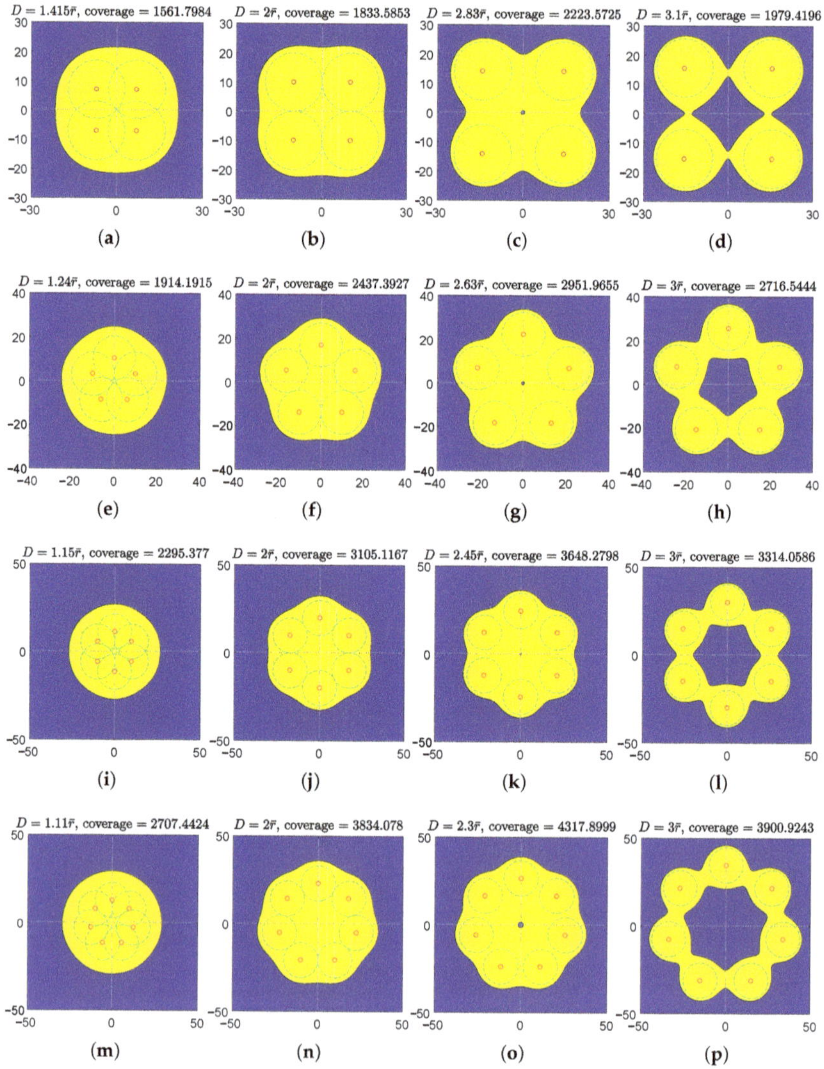

Figure 9. Illustration of effective coverages of n-node teams equipped with model $I(d) = \beta/d^2$ where $n = 4, 5, 6, 7$. The teams are in regular polygon formations. $\bar{r} = 10$. Each row represents an n value. In the first column (**a**,**e**,**i**,**m**), individual effective coverage disks of every other node (i.e., i and $i+2$) on the polygon vertices are tangent. In the second column (**b**,**f**,**j**,**n**), the individual effective coverage disks of every adjacent node pair are tangent. The third column (**c**,**g**,**k**,**o**) is the "max column", where teams' effective coverages are maximized. Each effectively covered region in the fourth column (**d**,**h**,**l**,**p**) has an effective coverage hole inside, while the region is still connected.

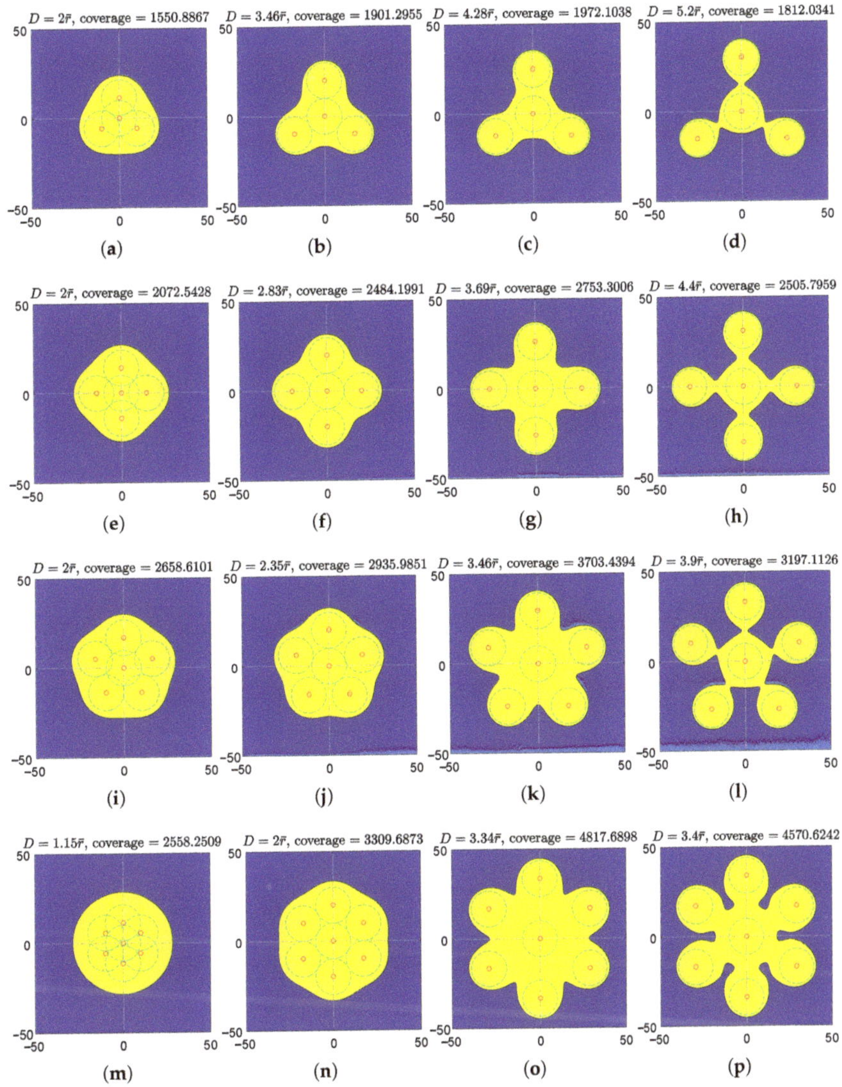

Figure 10. Illustration of effective coverages of n-node teams equipped with model $I(d) = \beta/d^2$, where $n = 4, 5, 6, 7$. The teams are in regular star formations. $\bar{r} = 10$. Each row represents an n value. In the first column, individual effective coverage disks of every adjacent node pair are tangent in (**a**,**e**,**i**), and individual effective coverage disks of every other node on the polygon vertices are tangent in (**m**). In the second column (**b**,**f**,**j**,**n**), the individual effective coverage disks of nodes on the vertices are tangent to that of the center node. The third column (**c**,**g**,**k**,**o**) is the "max column", where teams' effective coverages are maximized. The fourth column (**d**,**h**,**l**,**p**) shows the effective coverages at randomly picked separations while the region is still connected.

The effective coverages of a team and per node are plotted in Figure 12. From Figure 12a, it is seen that the team's effective coverage naturally increases with the number of nodes in both formations. It can be observed that with $n = 4, 5$, regular polygon formations generate larger maximum effective coverages than regular star formations; while with larger n values ($n = 6, 7$), the regular star formations instead generate larger maximum effective coverages than regular polygon formations. This trend is further revealed in subplot (c). From Figure 12b, it is found that for both regular polygon and regular star formations, the maximum effective coverage per node increases with the number of nodes n. This is further revealed in subplot (d), where we can see that the single-node contribution to effective coverage tends to increase slower when the number of nodes n increases. The optimal side lengths of both polygon and star formations decrease as the number of nodes n grows, which is shown in Figure 12e.

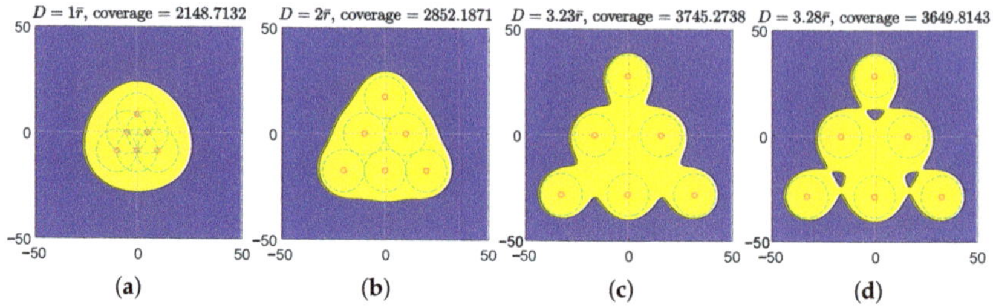

Figure 11. Illustration of effective coverages of n-node teams equipped with model $I(d) = \beta/d^2$, where $n = 6$. The teams are in equilateral triangle formations. $\bar{r} = 10$. (**a**) $D = \bar{r}$; (**b**) $D = 2\bar{r}$, individual effective coverage disks of adjacent nodes are tangent with each other; (**c**) the team's effective coverage maximized; (**d**) the effective coverage at randomly picked separation while the region is still connected.

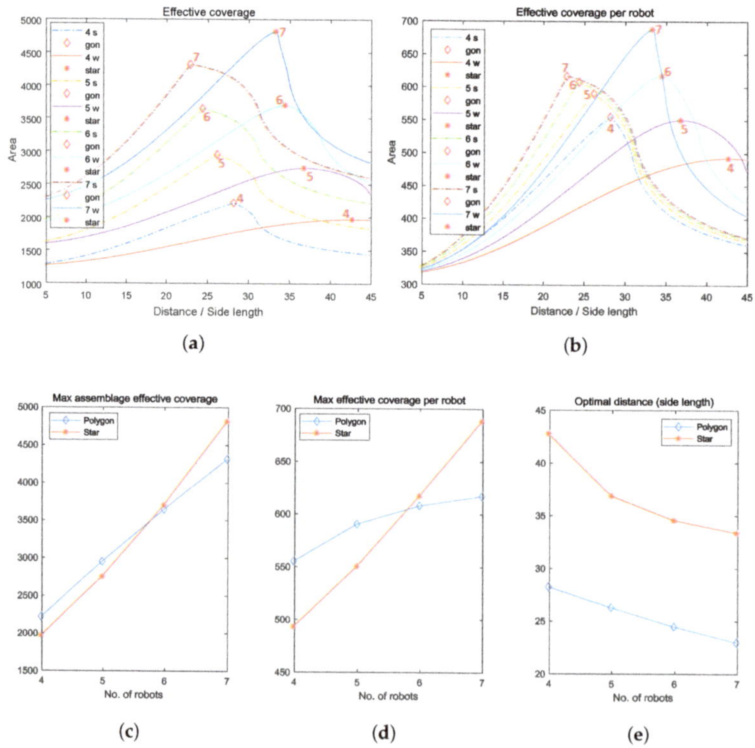

Figure 12. Plots of effective coverages of n-node teams equipped with model $I(d) = \beta/d^2$, where $n = 4, 5, 6, 7$. The team formations are in regular polygons ($\diamond$) and regular stars (*). $\bar{r} = 10$. (**a**) Team's effective coverage; (**b**) Effective coverage per node; (**c**) Maximum team's effective coverage vs. The number of nodes; (**d**) Maximum effective coverage per node vs. The number of nodes; (**e**) Optimal distance (polygon side length) vs. The number of nodes.

3.4. Comparison

To justify the effectiveness of our work in solving the given problem, a comparison with the existing binary model and the non-constant models in infinite planes (e.g., [5]) is conducted. By applying Equation (15) as the real sensing quality (confidence) function, the team's cooperative sensing quality at a point p is $\Phi(p) = 1 - \prod_{i=1}^{n}(1 - I(d_i)) = 1 - \prod_{i=1}^{n}\left(1 - \frac{\alpha}{d_i^{\lambda}+\alpha}\right)$. Let's examine $n = 4$ as an example. By using the same method with previous subsections, the optimal side length D_m and the largest effective coverage of the team are obtained and logged in Table 1. With the binary model ($\bar{r} = r$), the team in regular polygon formation achieves the maximum effective coverage at $D \geq 2r$ obviously; while to keep its connectedness, its D_m has to be $2r$.

From Table 1, it can be easily seen that our model obtains the closest optimal formations and their corresponding effective coverage values with the ground truth. This is because the mathematical expression of the binary model is too far from that of a real device. The models for solving the coverage problem of an infinite plane only consider the portion inside a polygon lattice of the infinite plane; the rest of the portion outside of the polygon lattice is considered as simple repetitions in other neighboring polygons, thus causing larger errors in approximation.

Table 1. Comparison of D_m and $\bar{C}$ values with $n = 4$.

α	Ground Truth	Our Convex Model	Binary Model	Infinite Plane Model (e.g., [5])
1	$2.835\bar{r}$, 2229.882	$2.84\bar{r}$, 2235.9219	$2\bar{r}$, 1256.6371	$2.845\bar{r}$, 3222.7623
3	$2.845\bar{r}$, 2241.6723	$2.86\bar{r}$, 2259.9843	$2\bar{r}$, 1256.6371	$2.87\bar{r}$, 3271.84
5	$2.855\bar{r}$, 2253.7023	$2.88\bar{r}$, 2283.6948	$2\bar{r}$, 1256.6371	$2.89\bar{r}$, 3317.76
7	$2.865\bar{r}$, 2265.4527	$2.9\bar{r}$, 2307.0858	$2\bar{r}$, 1256.6371	$2.915\bar{r}$, 3364
10	$2.88\bar{r}$, 2282.5741	$2.93\bar{r}$, 2342.5067	$2\bar{r}$, 1256.6371	$2.945\bar{r}$, 3436.362

4. Conclusions

In this article, the general properties of the effective coverage of an n-node team ($n \geq 3$) equipped with radial attenuation disk models following the superposition principle have been studied. For the 3-node team equipped with an unbounded convex model in the equilateral triangle formation, the general approach for finding the "last-connection" distance and its corresponding point coordinates are established. Example analyses and numerical simulations on both the unbounded convex model and the concave model have been conducted. For an n-node team ($n \geq 4$) in regular polygon formation where nodes are equipped with an unbounded convex model, it is proven that the circumcenter of the formation is the point with the weakest coverage intensity in the team's effectively covered region. Formations of regular polygons, regular stars, and an equilateral triangular tessellation (for six nodes) are simulated, through which the optimal formations are found. From the numerical result, for the 3-node team equipped with a concave model, the optimal distance is the individual effective coverage radius $\bar{r}$ when the threshold is high ($\bar{r}$ is small), and it falls between two times the individual effective coverage radius, the sum of the individual coverage radius, and the individual effective coverage radius ($r + \bar{r}$) when the individual effective coverage radius is roughly larger than 0.3 times the individual coverage radius. For 4- and 5-node teams, regular polygon formations have larger maximum effective coverages than their corresponding regular star formations, while when the number of nodes is larger than 6, the situation is just the opposite. For a 6-node team equipped with an unbounded convex model, the equilateral triangle formation is the optimal one with the largest maximum effective coverage among the three formations. For both the regular polygon and the regular star formations, the more the cooperative nodes, the larger the maximum effective coverage per node. The single-node contribution to the effective coverage of the team increases more slowly when the number of nodes increases. It is seen that the simulation results comply with the theoretical analysis and validate our propositions and theories. The comparison among our model, the binary model, and the method used for covering an infinite plane shows that our model is more accurate. This study provides useful insights and guides to the deployment of practical devices (e.g., sensors and other output devices) in coverage applications.

Author Contributions: Conceptualization, Y.-R.Y.; Methodology, Y.-R.Y.; Software, Y.-R.Y.; Validation, Y.-R.Y. and R.S.; Formal analysis, Y.-R.Y.; Investigation, Y.-R.Y.; Resources, Y.-R.Y.; Data curation, Y.-R.Y.; Writing—original draft, Y.-R.Y.; Writing—review & editing, Y.-R.Y.; Visualization, Y.-R.Y.; Project administration, Q.K. All authors have read and agreed to the published version of the manuscript.

Funding: This research received no external funding.

Institutional Review Board Statement: Not applicable.

Informed Consent Statement: Not applicable.

Data Availability Statement: No extra data other than those reported in the article is available.

Conflicts of Interest: The authors declare no conflict of interest.

Abbreviations

The following abbreviations are used in this manuscript:

WSN	Wireless Sensor Network

Appendix A. Sum of a Sine Function with Equidistant Phases

The following equation holds for $\theta \in \mathbb{R}, n \in \mathbb{Z}^+, n \geq 2$:

$$\sum_{i=1}^{n} \cos\left(\theta - \frac{2\pi i}{n}\right) = 0$$

There are three ways to prove this:

Appendix A.1. Proof by Contradiction Using a Vector Rotation Approach

Let's construct a group of n vectors $\cos\left(\theta - \frac{2\pi i}{n}\right)\mathbf{x} + \sin\left(\theta - \frac{2\pi i}{n}\right)\mathbf{y}$ for $i = 1, 2, \ldots, n$ with a common initial point at the origin. Assume the sum vector

$$\sum_{i=1}^{n} \left[\cos\left(\theta - \frac{2\pi i}{n}\right)\mathbf{x} + \sin\left(\theta - \frac{2\pi i}{n}\right)\mathbf{y}\right]$$

is a non-zero vector. When the entire vector group is rotated about the origin for $2\pi/n$, it becomes exactly the same as before rotation, but the non-zero sum vector is also rotated for $2\pi/n$. Hence, to be equal before and after the rotation, the sum vector must be a zero vector, i.e., $\sum_{i=1}^{n} \cos\left(\theta - \frac{2\pi i}{n}\right) = \sum_{i=1}^{n} \sin\left(\theta - \frac{2\pi i}{n}\right) = 0$ holds for any $\theta \in \mathbb{R}$.

Appendix A.2. Vieta's Formula Approach

We know that the ith root of equation $x^n = 1$ is $\cos\frac{2\pi i}{n} + j\sin\frac{2\pi i}{n}$. From Vieta's formulas of relations between roots and coefficients, we have

$$\sum_{i=1}^{n} \left(\cos\frac{2\pi i}{n} + j\sin\frac{2\pi i}{n}\right) = 0 \tag{A1}$$

which implies $\sum_{i=1}^{n} \cos\frac{2\pi i}{n} = \sum_{i=1}^{n} \sin\frac{2\pi i}{n} = 0$. Hence,

$$\sum_{i=1}^{n} \cos\left(\theta - \frac{2\pi i}{n}\right) = \cos\theta \sum_{i=1}^{n} \cos\frac{2\pi i}{n} + \sin\theta \sum_{i=1}^{n} \sin\frac{2\pi i}{n} = 0. \tag{A2}$$

Appendix A.3. Eular's Formula Approach

From Eular's formula, we have

$$\begin{aligned}
&\sum_{i=1}^{n}\cos\left(\theta-\frac{2\pi i}{n}\right)+j\sum_{i=1}^{n}\sin\left(\theta-\frac{2\pi i}{n}\right)\\
=&\sum_{i=1}^{n}e^{j\left(\theta-\frac{2\pi i}{n}\right)}\\
=&e^{j\theta}\sum_{i=1}^{n}e^{-j\frac{2\pi i}{n}}\\
=&e^{j\left(\theta-\frac{2\pi}{n}\right)}\frac{1-e^{-i2\pi}}{1-e^{-j\frac{2\pi}{n}}}\\
=&e^{j\left(\theta-\frac{2\pi}{n}\right)}\frac{1-1}{1-e^{-j\frac{2\pi}{n}}}\\
=&0.
\end{aligned} \tag{A3}$$

Hence, $\sum_{i=1}^{n}\cos\left(\theta-\frac{2\pi i}{n}\right)=\sum_{i=1}^{n}\sin\left(\theta-\frac{2\pi i}{n}\right)=0$ holds for any $\theta\in\mathbb{R}$.

Appendix B. Relation between First- and Second-Order Derivatives of a Convex Decreasing Function

For a convex model of coverage $I(d)$, $\forall R\in\mathbb{R}^{+}$, the following inequality always holds:

$$\left.\frac{\mathrm{d}^2I(d)}{\mathrm{d}d^2}\right|_{d=R}+\frac{1}{R}\left.\frac{\mathrm{d}I(d)}{\mathrm{d}d}\right|_{d=R}>0 \tag{A4}$$

Proof. Convex models of coverage possess the following properties:

- $I(d)>0$ on $d\in\mathbb{R}^{+}$
- $\frac{\mathrm{d}I(d)}{\mathrm{d}d}<0$ on $d\in\mathbb{R}^{+}$
- $\frac{\mathrm{d}^2I(d)}{\mathrm{d}d^2}>0$ on $d\in\mathbb{R}^{+}$
- $\lim\limits_{d\to+\infty}I(d)=0$
- $\lim\limits_{d\to+\infty}\frac{\mathrm{d}I(d)}{\mathrm{d}d}=0$

According to the above properties, we approximately draw the curve for the first derivative $\mathrm{d}I/\mathrm{d}d$ v.s. d, which entirely falls in the fourth quadrant and is asymptotic to d-axis, as shown in Figure A1a. Then, we flip the function image about the d-axis to obtain a new curve for $-\mathrm{d}I(d)/\mathrm{d}d$, as shown in Figure A1b. Next, we flip the new image again about the line $d=R$ to obtain the final curve for $-\mathrm{d}I(2R-d)/\mathrm{d}d$, as shown in Figure A1c. On the final curve, the derivative at $d=R$ equals that on the original curve:

$$-\left.\frac{\mathrm{d}^2I(2R-d)}{\mathrm{d}d^2}\right|_{d=R}=\left.\frac{\mathrm{d}^2I(d)}{\mathrm{d}d^2}\right|_{d=R}>0 \tag{A5}$$

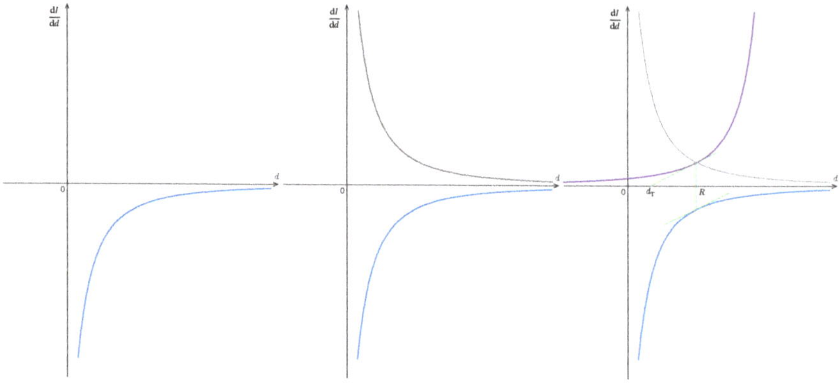

(**a**) (**b**) (**c**)

Figure A1. Plots of the first-order derivative of a convex model of coverage and its mirrors: (**a**) original first-order derivative of the convex model (blue), (**b**) mirrored about *d* axis (gray), and (**c**) further mirrored about $d = R$ (violet).

We draw the tangent line on curve $-\mathrm{d}I(2R-d)/\mathrm{d}d$ at $d = R$, which intersects d-axis at d_{T}, as shown in Figure A1. Since $\mathrm{d}I(d)/\mathrm{d}d$ is concave, $-\mathrm{d}I(2R-d)/\mathrm{d}d$ is convex, and $d_{\mathrm{T}} > 0$. So, we have

$$-\left.\frac{\mathrm{d}I(d)}{\mathrm{d}d}\right|_{d=R} = -\left.\frac{\mathrm{d}^2 I(2R-d)}{\mathrm{d}d^2}\right|_{d=R}(R-d_{\mathrm{T}}) < \left.\frac{\mathrm{d}^2 I(d)}{\mathrm{d}d^2}\right|_{d=R} R \tag{A6}$$

Therefore,

$$\frac{1}{R}\left.\frac{\mathrm{d}I(d)}{\mathrm{d}d}\right|_{d=R} + \left.\frac{\mathrm{d}^2 I(d)}{\mathrm{d}d^2}\right|_{d=R} > 0 \tag{A7}$$

□

References

1. Yáng, Y.R.; Chen, P.C. Effective Coverage of Two-Node Team with Radial Attenuation Models. *J. Intell. Robot. Syst.* **2021**, *103*, 1–13. [CrossRef]
2. Gupta, H.P.; Tyagi, P.K.; Singh, M.P. Regular Node Deployment for k-Coverage in m-Connected Wireless Networks. *IEEE Sens. J.* **2015**, *15*, 7126–7134. [CrossRef]
3. Sakai, K.; Sun, M.T.; Ku, W.S.; Lai, T.H.; Vasilakos, A.V. A Framework for the Optimal k-Coverage Deployment Patterns of Wireless Sensors. *IEEE Sens. J.* **2015**, *15*, 7273–7283. [CrossRef]
4. Wang, B.; Deng, X.; Liu, W.; Yang, L.T.; Chao, H. Confident information coverage in sensor networks for field reconstruction. *IEEE Wirel. Commun.* **2013**, *20*, 74–81. [CrossRef]
5. Zhu, J.; Wang, B. The optimal placement pattern for confident information coverage in wireless sensor networks. *IEEE Trans. Mob. Comput.* **2016**, *15*, 1022–1032. [CrossRef]
6. Yen, L.H.; Yu, C.W.; Cheng, Y.M. Expected k-coverage in wireless sensor networks. *Ad Hoc Netw.* **2006**, *4*, 636–650. [CrossRef]
7. Wang, B. Coverage problems in sensor networks: A survey. *ACM Comput. Surv.* **2011**, *43*, 1–53. [CrossRef]
8. Yáng, Y.R.; Chen, P.C.Y. Navigation of Robot-Formation via Overlap-Induced Equilibrium Points in Uncertain Environments. *IEEE Trans. Autom. Sci. Eng.* **2022**, 1–12. [CrossRef]
9. Bai, X.; Yun, Z.; Xuan, D.; Lai, T.H.; Jia, W. Optimal patterns for four-connectivity and full coverage in wireless sensor networks. *IEEE Trans. Mob. Comput.* **2010**, *9*, 435–448.
10. Yun, Z.; Bai, X.; Xuan, D.; Lai, T.H.; Jia, W. Optimal deployment patterns for full coverage and k-Connectivity ($k \leq 6$) wireless sensor networks. *IEEE/ACM Trans. Netw.* **2010**, *18*, 934–947.
11. Zhang, C.; Bai, X.; Teng, J.; Xuan, D.; Jia, W. Constructing low-connectivity and full-coverage three dimensional sensor networks. *IEEE J. Sel. Areas Commun.* **2010**, *28*, 984–993. [CrossRef]
12. Ranieri, J.; Chebira, A.; Vetterli, M. Near-Optimal Sensor Placement for Linear Inverse Problems. *IEEE Trans. Signal Process.* **2014**, *62*, 1135–1146. [CrossRef]
13. Lin, F.; Chiu, P. A near-optimal sensor placement algorithm to achieve complete coverage-discrimination in sensor networks. *IEEE Commun. Lett.* **2005**, *9*, 43–45. [CrossRef]

14. Wang, B.; Xu, H.; Liu, W.; Yang, L.T. The Optimal Node Placement for Long Belt Coverage in Wireless Networks. *IEEE Trans. Comput.* **2015**, *64*, 587–592. [CrossRef]
15. Li, L.; Chen, H. UAV Enhanced Target-Barrier Coverage Algorithm for Wireless Sensor Networks Based on Reinforcement Learning. *Sensors* **2022**, *22*, 6381. [CrossRef] [PubMed]
16. Akram, J.; Munawar, H.S.; Kouzani, A.Z.; Ma mud, M.A.P. Using Adaptive Sensors for Optimised Target Coverage in Wireless Senso Networks. *Sensors* **2022**, *22*, 1083. [CrossRef] [PubMed]
17. Tirandazi, P.; Rahiminasab, A.; Ebadi, M. An efficient coverage and connectivity algorithm based on mobile robots for wireless sensor networks. *J. Ambient. Intell. Humaniz. Comput.* **2022**, 1–23. [CrossRef]
18. Wang, B.; Wang, W.; Srinivasan, V.; Chua, K.C. Information coverage for wireless sensor networks. *IEEE Commun. Lett.* **2005**, *9*, 967–969. [CrossRef]
19. Deng, X.; Tang, Z.; Yang, L.T.; Lin, M.; Wang, B. Confident Information Coverage Hole Healing in Hybrid Industrial Wireless Sensor Networks. *IEEE Trans. Ind. Inform.* **2018**, *14*, 2220–2229. [CrossRef]
20. Dai, L.; Wang, B.; Yang, L.T.; Deng, X.; Yi, L. A Nature-Inspired Node Deployment Strategy for Connected Confident Information Coverage in Industrial Internet of Things. *IEEE Internet Things J.* **2019**, *6*, 9217–9225. [CrossRef]
21. Meng, S.; Kan, Z. Deep Reinforcement Learning-Based Effective Coverage Control With Connectivity Constraints. *IEEE Control Syst. Lett.* **2022**, *6*, 283–288. [CrossRef]
22. Radermacher, R. Robotic Personal Conditioning Device. University of Maryland. 2018. Available online: https://arpa-e.energy.gov/technologies/projects/robotic-personal-conditioning-device (accessed on 20 July 2020).

sensors

Article

Prehensile and Non-Prehensile Robotic Pick-and-Place of Objects in Clutter Using Deep Reinforcement Learning

Muhammad Babar Imtiaz *, Yuansong Qiao and Brian Lee

Software Research Institute, Technological University of the Shannon: Midland Midwest, N37 HD68 Athlone, Ireland
* Correspondence: b.imtiaz@research.ait.ie

Abstract: In this study, we develop a framework for an intelligent and self-supervised industrial pick-and-place operation for cluttered environments. Our target is to have the agent learn to perform prehensile and non-prehensile robotic manipulations to improve the efficiency and throughput of the pick-and-place task. To achieve this target, we specify the problem as a Markov decision process (MDP) and deploy a deep reinforcement learning (RL) temporal difference model-free algorithm known as the deep Q-network (DQN). We consider three actions in our MDP; one is 'grasping' from the prehensile manipulation category and the other two are 'left-slide' and 'right-slide' from the non-prehensile manipulation category. Our DQN is composed of three fully convolutional networks (FCN) based on the memory-efficient architecture of DenseNet-121 which are trained together without causing any bottleneck situations. Each FCN corresponds to each discrete action and outputs a pixel-wise map of affordances for the relevant action. Rewards are allocated after every forward pass and backpropagation is carried out for weight tuning in the corresponding FCN. In this manner, non-prehensile manipulations are learnt which can, in turn, lead to possible successful prehensile manipulations in the near future and vice versa, thus increasing the efficiency and throughput of the pick-and-place task. The Results section shows performance comparisons of our approach to a baseline deep learning approach and a ResNet architecture-based approach, along with very promising test results at varying clutter densities across a range of complex scenario test cases.

Keywords: prehensile; non-prehensile; robotic manipulation; Markov decision process; deep reinforcement learning; deep Q-network; fully convolutional network; DenseNet-121

Citation: Imtiaz, M.B.; Qiao, Y.; Lee, B. Prehensile and Non-Prehensile Robotic Pick-and-Place of Objects in Clutter Using Deep Reinforcement Learning. *Sensors* **2023**, *23*, 1513. https://doi.org/10.3390/s23031513

Academic Editors: Xin Zhao, Mingzhu Sun and Qili Zhao

Received: 9 December 2022
Revised: 24 January 2023
Accepted: 27 January 2023
Published: 29 January 2023

1. Introduction

The term Industry 4.0 has been around for more than a decade now. As the fourth industrial revolution was all about automation, industry became familiar with various cutting-edge technologies that were crucial to achieving the goal of automation, such as digital twins, cyber–physical systems, smart manufacturing, Internet of Things (IoT), etc.

The deployment of robotics in the industry, in one way or another, can be traced back almost 50 years. Usually, robotic manipulations can be divided into two kinds: prehensile and non-prehensile manipulation. Prehensile motion [1] involves grasping or capturing, whereas non-prehensile motion [2] is where propelling or pushing is involved instead of grasping. An example of prehensile and non-prehensile motion is shown in Figure 1 from [3]. Well-learned coordination between these two robotic manipulations can lead to higher success rates of pick-and-place; for instance, non-prehensile manipulation can create space between tightly packed objects, thus making grasping possible.

On the other hand, prehensile manipulation will pick objects and facilitate collision-free and more accurate pushing. A substantial amount of work has been conducted on prehensile and non-prehensile robotic manipulations in the past, but mostly, they have been researched independently of each other [4]. There is still a considerable research gap and potential for exploration in the area of the combination of these robotic manipulations.

Non-prehensile manipulation has been primarily used to accurately deal with the poses of the objects [2]. In a combined approach involving grasping and pushing, pushing goals are quite flexible; for example, they can create space between objects, change the pose of objects, and crack clutter [5]. This flexibility factor was not addressed in different model-based [6–8] and data-driven [9–11] approaches in the past. Model-based approaches are those which are based on a static model, whereas data-driven approaches are those which involve any kind of machine learning through data. In the area of prehensile manipulation, there have been different approaches to learning successful grasping policies, some based on optimizing affordances by learning from previous experiences [12–14] and some learnt through grasp stability measures [15–17]. These approaches do not address the planning or learning of sequences of prehensile and non-prehensile manipulation actions to maximize the success rate of grasping and pushing. There are some hard-coded techniques that are dependent on domain knowledge for implementing single-action push–grasp but, obviously, they have very limited scope [2].

Figure 1. An example of prehensile and non-prehensile manipulations.

In this study, we present a framework in which the agent learns to perform a sequence of prehensile and non-prehensile robotic manipulations through model-free temporal difference off-policy deep reinforcement learning (Q-learning). This approach does not require any kind of prior domain knowledge or geometrical pose information like some previous approaches, which will be further discussed in the next section. The major objectives achieved in this study are as follows:

- We designed a self-supervised deep reinforcement learning-based data-driven framework. We deploy a memory-efficient joint network in this framework, unlike the approaches in the past. This framework successfully enables the agents to learn and perform a sequence of prehensile and non-prehensile operations together on regular- and irregular-shaped objects without causing any bottleneck situations, unlike preceding work. This learning is achieved through trial and error.
- The sliding or pushing operations are learnt in a bidirectional manner (left and right), which then further enables grasping operation. This learning of bidirectional non-prehensile manipulation directly elevates the progress of unidirectional non-prehensile manipulation. This contrasts with approaches in the past using hard-coded techniques, domain-based knowledge such as heuristics, or unidirectional prehensile manipulations [2].
- Our agents learn and converge to the optimal policies through a joint framework of end-to-end memory-efficient deep networks. In end-to-end networks, the model learns all the steps between the initial input phase and the final output result; therefore, memory management is essential to avoid bottleneck situations while using minimal resources. This joint framework takes the visual state of the workspace as an input, and in return, outputs the Q-values depicting the expected return for three different potential actions: grasping, sliding left, and sliding right. Once the Q-values are received, the action associated with the highest Q-value is selected as it predicts the maximum expected future reward. This whole procedure will be explained in detail in Section 3 but even

this brief overview clearly shows that our approach contrasts with the approaches of the past, where each and every object was considered individually and hand-designed features and heuristics were crucial for the action selection part [18].

In light of the abovementioned research objectives of this paper, the following three elements represent the novelty of this proposed research framework:

- (i) Efficient memory management: this paper avoids any kind of bottleneck situation at the GPU end through the deployment of a memory-efficient variant of DenseNet-121 [19], unlike the preceding works.
- (ii) Bi-directional Sliding operations: learns bi-directional non-prehensile manipulation (left and right direction sliding) in contrast to unidirectional pushing in the existing approaches.
- (iii) Additional overhead avoidance: our framework avoids any kinds of additional overheads incurred by previous approaches due to the use of object segmentation and singulation techniques.

There are some existing mini production line setups, i.e., robotic arms and conveyor belts [20,21], available on the market for pick-and-place tasks but they are completely hard-coded and do not have any learning capability unless linked with a learning-based frame such as the one proposed in this paper.

With the approach described in the above-mentioned manner, the problem can be seen as a labeling problem where every pixel is labeled. Every pixel in the current state correlates with the prehensile and non-prehensile robotic manipulations. These manipulations are carried out at the three-dimensional location of the pixel in the current state. In the case of prehensile motion, the pixel location will be considered as the center point of grasping, whereas in the case of non-prehensile robotic manipulation, it will be taken as the initiation point of the left or right-slide. We train a fully convolutional network and supply it with the RGB-D images of the current scene of the workspace and obtain a pixel-wise prediction table as the output. This pixel-wise prediction table helps the agent select the appropriate pixel for robotic manipulation. These pixel-wise predictions procedures, also known as fully convolutional action-value functions [12], provide us with the expected future reward values for all the pixels whereby the pixel predicted to have the highest expected reward is chosen for robotic manipulation on the three-dimensional surface. These pixel-wise action-value functions make our agent learn and perform three different robotic manipulation policies after just a few hours of training.

In summary, our overall contribution is a data-driven framework that involves interlinked prehensile and non-prehensile robotic manipulations learned and performed through end-to-end deep networks based on the concept of pixel-wise parameterization. The main advantage of this work over others is that it deploys prehensile and non-prehensile manipulations in a sequential manner such that it outputs a high success rate for pick-and-place of regular and irregular objects in cluttered environments. Clutter can be seen as a randomly closely packed group of regular- and irregular-shaped objects. Cluttered objects are a common scenario in bin picking jobs in the industry. Similar to simple or sparse pick-and-place, bin picking where objects are to be picked from clutter has applications for implementation in all types of sectors such pharmacy, automotive, logistics and many others [22]. A few examples are: the world's largest shock absorber manufacturer in Indiana, US requiring the picking of steel metal cylinders from clutter and their placement for further processing [23]; a leading door manufacturing company in the US requiring the packing of two different plastic parts after picking them from randomly organized clutter in two separate bins [23]; and the Italian sports gun manufacturer Benelli requiring small parts in clutter to be picked from bins and kitted in plastic boxes every day at its factory [23]. Responding to this worldwide demand, several research approaches have been presented for pick-and-place from clutter. The popular competition the Amazon Picking Challenge, which took place from 2015 to 2017 [24], was also focused on bin picking owing to its importance in industrial operations. Pixel-wise parameterization achieved through end-to-end deep networks is the key concept in our approach and will be explained in later sections. We present the results of various experimentations performed in a simulated environment designed using a V-REP simulator.

2. Related Work

Grasp planning's history can be traced back to the beginning of the robotic manipulation era. Some studies consider the classification of grasps on the basis of their potential to restrict or limit the mobility of the object [25], while some other studies involve the modeling of contact forces and their opposing relation to grasping [4,26]. Some approaches work with pre-computed grasp poses at runtime by linking them with the point cloud data being used for object pose estimation [27]. These pre-generated grasps for translating the abovementioned approaches to real-life environments are usually created from databases of 3D object models [28]. A such, they require detailed knowledge beforehand, such as the poses, shapes, contact points, etc., of the objects under consideration, which is almost impossible to guarantee when dealing with novel objects in real-life environments. Aside from model-driven studies, various data-driven approaches have also recently emerged. Several studies have been carried out to train model-independent grasping policies in which an agent learns to detect potential grasps without having any prior object knowledge of factors such as the object's pose, shape, contact points, etc. These policies are trained by learning the visual features instead [5,12,13,15,16,29]. Studies such as [5] present the idea of using pre-trained models on other robotic manipulations to decrease the learning effort while learning the target robotic manipulation (for instance, a model pre-trained on poking trying to learn grasping). In [12], deep networks were used to predict pixel-based affordances in order to make agents learn grasping policies. Similar to these studies, the approach presented in this paper is model-independent but involves the learning of a sequence of prehensile and non-prehensile robotic manipulations, leading to an efficient pick-and-place agent.

Non-prehensile motion can be dated back to the 1970s [30]. Most approaches presented in the early days tackled the task by modeling it based on frictional forces [7,8]. While these approaches motivated the research in this area, their assumptions while modeling the task, such as variability and the non-uniform distribution of friction, were not good enough to hold in real-world settings [10]. This caused these kinds of non-prehensile robotic techniques to fail in a real-world setting by making inaccurate predictions. Data-driven approaches to learn and perform non-prehensile robotic manipulation have also been presented [31]. Most of these data-driven approaches focused on training the agents to consider one object at a time. It is still a complex challenge to model clutter-based non-prehensile manipulation and develop solutions that can be proven effective in real-world scenarios.

Significant benefits can be obtained by combining both prehensile and non-prehensile approaches. The study presented in [2] is a pioneering model-driven approach to performing a push–grasp operation, a robotic manipulation in which grasping is performed within non-prehensile manipulations. With this method, grasping chances are increased. The study targets clutter-based workspace and it also focuses on achieving the ability to move objects in clutter. However, this whole technique requires domain knowledge beforehand, i.e., it is completely handcrafted. In contrast, the data-driven approach in this paper requires no domain knowledge while learning through self-supervision. There are some other approaches where grasping policies are already known and the task is to move objects using non-prehensile robotic manipulation to certain locations. These certain locations are locations where grasping can be performed according to the grasping policies known beforehand [9,32]. A reinforcement learning-based approach has been presented in [18] where the agent is trained to learn policies based on handcrafted features. The agent must choose between prehensile and non-prehensile manipulation in order to perform a pick-and-place task in a cluttered environment. The flow of their approach can be seen as first performing segmentation on RGB-D images received from the sensor. Through the segmentation, objects are identified. Once objects have been identified, multiple instances of actions are sampled (prehensile and non-prehensile manipulation) for each object. For each sampled instance, hand-crafted features are extracted and, finally, an action instance is performed that is predicted to have the maximum expected reward. The limitations to this approach are that it is trained for mostly convex objects only. The non-prehensile policy or pushing policy involved requires handcrafted features such as prediction of the motion of the pushed object. This prediction is made with the help of a simulator. The future effect of this pushing policy on the prehensile manipulation also needs to be known. In [33], the authors present two linear push policies to push objects away from the walls in bin-picking. Singulation techniques through reinforcement learning have been also presented but are limited to regular-shaped objects and unidirectional pushes [34,35]. Some approaches in the recent past have also been presented where pixel-wise-parameterization techniques have been used for the robotic manipulation of only regular-shaped objects, but mostly, they did not address non-prehensile motion [36] or movement in a limited and unidirectional manner [12,37,38]. In such approaches, memory management is a key limitation to be addressed. In contrast, the approach presented in this paper involves policies that are trained with end-to-end memory-efficient deep networks through pixel-wise-parameterization for prehensile and bidirectional non-prehensile robotic manipulation. These policies do not require or assume any kind of object knowledge of factors such as pose, shape, contact points, etc. Our results show that our approach performs successfully in various test cases involving many regular and irregular objects in clutter.

3. Methodology

In our framework, the agent learned to perform picking of regular and irregular-shaped objects in clutter from a conveyor belt and to place them in a bin with the help of a series of prehensile and non-prehensile robotic manipulation operations. In reinforcement learning, an agent is the entity that uses a policy to maximize the expected return gained from transitioning between states of the environment. In our case, the robotic arm UR5 can be seen later as an agent interacting with the environment, which is the workspace. We used reinforcement learning where a policy is learnt to solve the task through sequential decision-making. As this is a learning-based data-driven framework, the success lies in learning to predict the most suitable pixel of the workspace for the grasping and sliding actions. The secondary success, as shown by the results, is the successful learning of synchronization between the prehensile (grasping) and non-prehensile (sliding) actions. The RL agent learns to perform such non-prehensile manipulations which lead to successfully rewarding prehensile manipulations in the near future and vice versa, whereas the grasping pose is considered static in our work, always from the top angle, and robotic arm motion planning is dynamic and controlled by the Open Motion Planning Library (OMPL) built into the V-REP simulator. The overall success of the framework was calculated by measuring the attempts taken to clear the workspace of all the objects, which could only be achieved efficiently if the agent was able to learn the abovementioned factors and the policy converged successfully. According to the design of the problem, an MDP was formulated. The basic elements of the MDP are as follows:

- $s \in S$ (RGB-D heightmap), where RGB-D heightmap represents the heightmap generated from the current scene of the workspace at any time instance. In this approach, we used vision sensors to obtain the view of the current state of the workspace. Two types of vision sensors were used in the environment: an orthographic projection vision sensor and a perspective projection vision sensor. The basic difference between these two is their field of view. Orthographic projection vision sensors have a rectangular projection view, whereas perspective projection vision sensors have a trapezoidal field of view. Figure 2 shows the location and fields of view of these vision sensors in our designed simulation setup. We captured RGB-D images through these fixed position vision sensors and aggregation was performed to gather maximum RGB and depth information. Basically, an RGB-D image constitutes an RGB image and its corresponding depth image. A depth image contains information relating to the distance of the surfaces of scene objects from a viewpoint, which, in our case, was the height-from-bottom information. While an RGB image has three channels (red, green, and blue), the depth image has one channel in which each pixel is in relation to the distance between the image plane and the corresponding object in the RGB image [39]. From these RGB-D images, we created RGB-D heightmaps. We followed the method of heightmap computation from the RGB-D images used in [12] where the RGB-D data are projected onto a 3D point cloud, which is then orthographically back-projected upwards in the gravity direction, leading to a heightmap representation of the workspace with both color (RGB) and height-from-bottom (D) channels. Thus, each RGB-D heightmap representation was taken as the state s_t at time t, as shown in Figure 3. The RGB-D heightmaps created through this process comprised 224 × 224 pixels—the size of our workspace. Each pixel in the heightmap represents a unique three-dimensional location in the physical workspace.

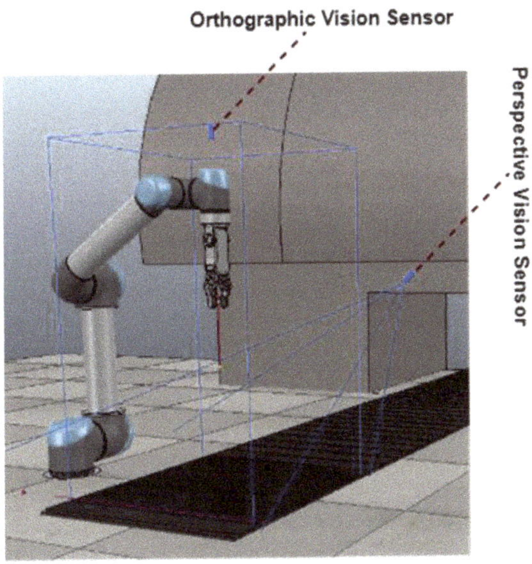

Figure 2. Location and field views of vision sensors in the simulated setup.

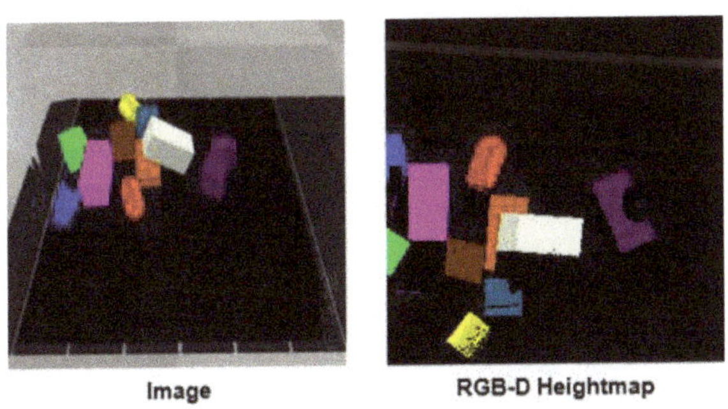

Figure 3. A sample of current view image from the workspace at timestep t and its generated RGB-D 224 × 224 pixel heightmap representation of the state s_t.

- a ∈ A: (grasp, left-slide, and right-slide), where grasp, left-slide, and right-slide are three unique available actions in this MDP. We designed each robotic manipulation as a distinct action, i.e., grasping (prehensile), left-slide and right-slide (non-prehensile). The motive behind choosing actions from both categories is that non-prehensile manipulations will assist and lead to prehensile manipulation. Each action a_t, at a given time t, will be performed at a three-dimensional location p' corresponding to a pixel p of the RGB-D heightmap or state s_t. This formulation can be seen as $p \rightarrow p' \in s_t$, where p' is the mid-point of the grasping. As we have top-mounted vision sensors, our grasping will be carried out from above using an RG2 gripper [40], which is a parallel-jaw grasper that went approximately 2 cm deeper than the p''s z coordinate, as shown in Figure 4, for a firm grasp. On the other hand, if left-slide or right-slide is to be performed, p' is the three-dimensional location that will serve as the starting point of an approximately 8 cm linear straight push with the tip of the closed parallel-jaw grasper in the left or right direction of the gripper, respectively, as shown in Figure 4. The left-slide and right-slide actions will be performed with reference to the robotic arm and the red and blue arrows show the direction of the right-slide and left-slide, respectively, in Figure 4. For instance, if we imagine ourselves as the robotic arm, our right hand points in the direction of the right-slide and our left hand points in the direction of the left-slide. These grasping and sliding actions can be seen in Figure 5, where Frame 1 shows the grasping action and, upon close inspection, Frames 2 to 6 show the sliding action where the green cuboid object is being slid/pushed by the robotic gripper. To accurately assess the sliding action, one can notice the distance between the green cuboid object and the brown triangular object reducing frame by frame, especially from Frames 4 to 6.

- r ∈ R (1, 0.8, 0.5, 0), where 1, 0.8, 0.5, and 0 are the unique rewards in our rewarding scheme. A reward scheme in a reinforcement learning approach is designed based on possible outcomes. In our approach, we considered four possible outcomes. Firstly, the object is grasped and placed at the designated location successfully; secondly, the object is grasped successfully but falls during displacement to the designated location (in the basket); thirdly, a successful left- or right-slide is performed; and finally, any other outcome occurs. For the first outcome, the agent is awarded 1; for the second outcome, which is verified by an infrared sensor between the gripper jaws, the agent is awarded 0.8; and for the third outcome, verified by finding changes in the workspace by comparing current and previous heightmaps, the agent is awarded 0.5. Any other outcome is considered a complete failure, so the agent is awarded 0 as a reward.

According to the flow of our designed MDP, our agent, i.e., the robotic arm, at any given state s_t at time t, according to the policy π, can choose and execute an action a_t from the list of available actions, ending up in a new state s_{t+1}. It then receives a reward $r_{at}(s_t, s_{t+1})$, meaning a reward is earned after performing action a at time t during the transition from state s_t to state s_{t+1}. The goal of this task is to discover the optimal policy π^* through which we can achieve the goal of maximization of the expected rewards in future, given by Equation (1) as follows:

$$R_t = \Sigma_{i=t}^{\infty} \gamma R_{ai}(s_i, s_{i+1}) \quad (1)$$

where γ denotes the discount factor, usually ranging between 0 and 1, thus making the agent learn at a faster pace. In this paper, we designed our approach based on an off-policy model-free temporal difference learning algorithm known as Q-learning. We aimed to achieve convergence of a policy π, which is greedy in nature and performs action selection based on the maximization of the action–value function, also known as the Q-function. A Q-function, which can be described as $Q_\pi(s_t, a_t)$, is actually our tool that predicts the expected reward for performing an action a_t while being at state s_t at any time t. We can formally see our Q-function working in Equation (2).

$$Q(s_t, a_t) \leftarrow Q(s_t, a_t) + \alpha\left[R_{at}(s_t, s_{t+1}) + \gamma Q(s_{t+1}, argmax(Q(s_{t+1}, a_t'))) - Q(s_t, a_t))\right] \quad (2)$$

where α denotes the learning rate and a' denotes the list of all available actions.

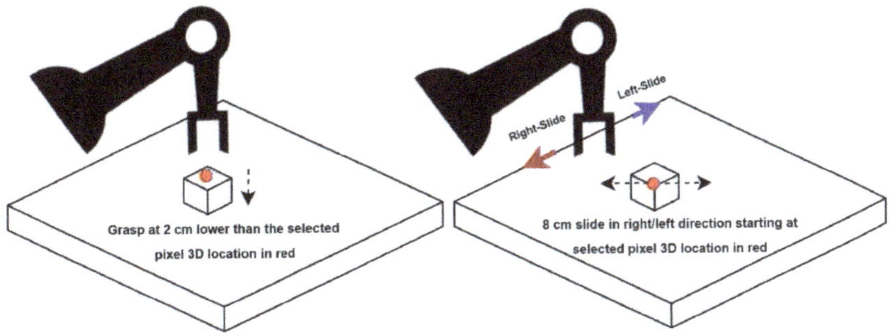

Figure 4. Grasp action (**left**) and left-slide and right-slide actions (**right**).

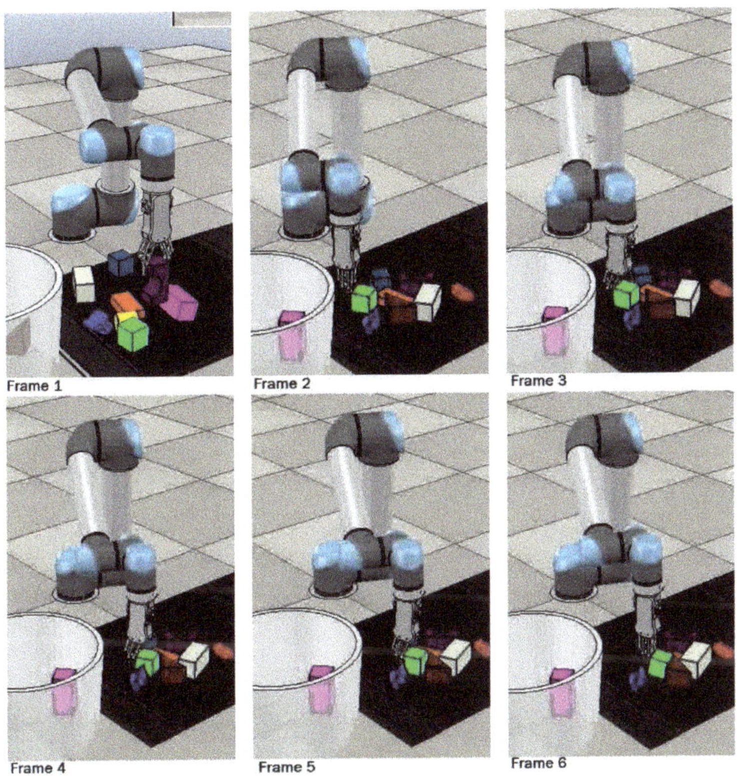

Figure 5. Frame 1 shows the grasping action whereas Frame 2 to Frame 6 show the sliding action being performed on the green square object.

3.1. Q-Function Design

We used deep Q-networks for Q-function approximation to obtain the state-action values that predict the expected future rewards for carrying out an action a_t at a state s_t [41,42]. The Q-function that we designed is a combination of feed-forward fully convolutional networks (FCNs) [43]. For more efficient learning of features, we created a distinct FCN for each of the available actions, which are denoted as $\{FCN_{Grasp}, FCN_{LeftSlide}, FCN_{RightSlide}\}$. Each of these FCNs is supplied with state s, which is represented as a heightmap, processes it, and generates a pixel-wise map as the output. The values in this map are the Q-values. As mentioned earlier, the state is represented by a heightmap of 224 × 224 pixels in accordance with the workspace size and the dense pixel-wise map generated by the FCN as the output is also of the same 224 × 224 size. Each value in this dense pixel-wise map corresponds to each corresponding pixel in the heightmap representation of the state. Each value in this pixel-wise map is a prediction of the expected future reward the agent will earn after performing the corresponding action at p', with the three-dimensional location corresponding to the pixel p. All three FCNs $\{FCN_{grasp},\ FCN_{LeftSlide},\ FCN_{RightSlide}\}$ have the same architecture, i.e., DenseNet-121 [44], whose general architecture is presented in Figure 6.

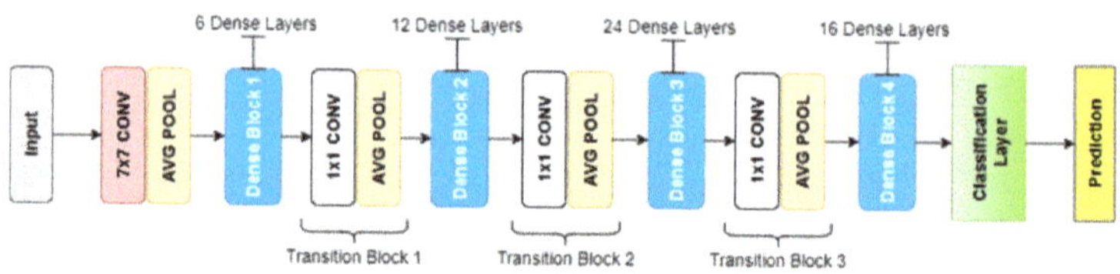

Figure 6. DenseNet-121 general architecture.

The main difference between other traditional convolutional networks and DenseNet is the number of connections. A traditional convolutional network comprising L layers has L connections, which means every layer is connected to its immediate successor layer; on the other hand, DenseNet comprises $L(L+1)/2$ connections, where any layer will obtain the feature maps from all its predecessor layers as inputs and will supply its own feature map to all its successor layers as their input. This is mainly the reason behind adopting the DenseNet architecture in our approach. With the help of DenseNet, feature reuse is enabled, feature propagation is made better, the vanishing gradient problem is addressed, and parameter reduction is achieved. However, at the same time, all these benefits lead to quadratic growth of the features as the network becomes deeper and deeper, which results in choking of whole network and the formation of a bottleneck scenario. Therefore, we addressed this issue in a different manner, which will be stated in future sections. DenseNet-121 was used in two manners: trained from scratch and also pre-trained on ImageNet [45]. The structure of DenseNet-121 was extended with two additional 1 × 1 convolutional layers along with rectified linear unit (ReLU) activation function [46] and batch normalization [47] procedures. We used the multimodal approach shown in Figure 7 as in [12], and within an FCN, we created two trunks of DenseNet-121; one was supplied with color channels (RGB) from the heightmap representation of the state, and the other trunk was supplied with clones across the three-channel depth (DDD) of the heightmap, which was normalized by subtracting the mean and dividing by the standard deviation. This cloning of depth across channels was performed to exploit the weights pre-trained on ImageNet RGB images. Features learned through both these trunks were concatenated and supplied to the extended 1 × 1 convolutional layers.

To enhance the learning of features for robotic manipulations, we brought in the rotation of heightmaps. Initially, we rotated the input heightmaps by 90 degrees, leading to 4 rotated heightmaps in total. Later, we experimented with rotating heightmaps at 45 and 22.5 degrees, leading to 8 and 16 rotated heightmaps in total, respectively. Figure 7 shows all three rotation modes; the first row shows the 90-degree rotations, adding the second row shows the 45-degree rotation, and all the rows together show the 22.5-degree rotations. The results show that features were learn better when heightmaps are rotated at 22.5 degrees, generating 16 heightmaps, so that the factor of different orientations is handled well. In this way, each FCN is supplied with 16 rotations of heightmaps and, in turn, outputs 16 dense pixel-wise Q-values maps. So, the total output generated from all the three FCNs $\{FCN_{Grasp},\ FCN_{LeftSlide},\ FCN_{RightSlide}\}$ is 48 dense pixel-wise maps, out of which 16 are for grasping, 16 are for the left-slide, and 16 are for right-slide. Simply put, the 48 maps are an extension of the 16-rotation scheme instead of being two different schemes. Processing was carried out for all three actions (grasp, left-slide, and right-slide) at parallel; each action's processing obtained 16 rotated pictures, and collectively, they amounted to 48 (16 + 16 + 16). In other words, when we consider a single action, it is 16 inputs and 16 outputs, but when we collectively consider all three action, it becomes 48 inputs and 48 outputs. In all these Q-values from the total 48 dense pixel-wise maps, the maximum one is selected for the particular action corresponding to the FCN it is generated from. Action was performed on p', the three-dimensional location of the pixel p whose corresponding maximum Q-value was selected. This is depicted as in Equation (3).

$$max\ Q(s_t, a_t) = max\Big(FCN_{Grasp}(s_t), FCN_{LeftSlide}(s_t), FCN_{RightSlide}(s_t)\Big) \tag{3}$$

In our pixel-wise affordances approach, these FCNs were used as Q-function approximators. FCNs have proven their worth in pixel-wise scenarios through efficient computation mechanisms. In each forward pass in our network, 2,408,448 Q-values are computed in a time-efficient manner. This total number of Q-values is calculated as each pixel-wise map is of 224 × 224 size, and such 48 pixel-wise maps are generated in every forward pass (224 × 224 × 48). The convergence of our policies was achieved with a lower training data requirement due to this pixel-wise parameterization and the involvement of rotation functionality (helping with learning various orientations), thus enabling the agent to learn at the maximum rate.

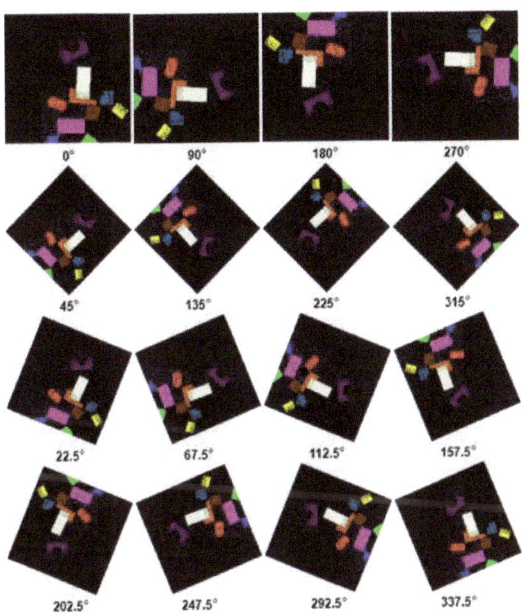

Figure 7. RGB-D heightmap rotations at 90 degrees (first row), 45 degrees (first and second row), and 22.5 degrees (all four rows).

3.2. Training

In each iteration, after receiving RGB-D images from the vision sensors, the heightmap representations are generated as the current state. The rotation function, shown in Figure 7, is then performed on the heightmaps, the rotated heightmaps are fed to the network, and pixel-wise maps are generated as outputs in return, as shown in Figure 8. As the same workspace state is being supplied, all three FCNs' outputs are shown on the same workspace image, but the difference or the actual output is the RED dot in all three output pictures labeled as the best location for each action. For instance, the top image shows the best location (red dot on the left edge of the white rectangular object) for the grasp action according to the 16 pixel-wise Q-value maps generated by its FCN_{Grasp}. The middle image shows the best location (red dot on almost the lower corner of the orange rectangular object) for the left-slide action according to the 16 pixel-wise Q-values maps generated by its $FCN_{LeftSlide}$. The bottom image shows the best location (red dot on the upper corner of the orange rectangular object) for the right-slide action according to the 16 pixel-wise Q-value maps generated by its $FCN_{RightSlide}$. The highest Q-value of these all three pixels (locations) will be selected for the corresponding action. These pixel-wise maps can be visualized in heatmap format as shown in Figure 9, where hot regions depict higher Q-values. For the backpropagation, we used the Huber loss function [48], which can be considered as a trade-off between the mean square error (MSE) and mean absolute error (MAE). In our backpropagation scheme, gradient passing was only performed through a particular FCN and pixel p, which gave the maximum Q-value for action selection, whereas 0 loss was backpropagated for all the remaining pixels. FCNs were trained using stochastic gradient descent with momentum [49], which is considered more efficient and rapid compared to traditional stochastic gradient descent (SGD). The traditional SGD calculates noisy derivatives of loss, whereas the SGD with a momentum variant involves exponentially weighted averages. The traditional SGD is outperformed by the momentum variant because these exponentially weighted averages are much closer to the actual loss compared to the SGD's noisy values. Therefore, an SGD with momentum helps gradient vectors to move in the right direction, leading to faster convergence. While training our network, the values of momentum, learning rate, and weight decay were set at 0.9, 1×10^{-2}, and 1×10^{-3}, respectively. In our training phase, we also used experience replay, where the agent keeps a record and reuses its experiences from the past. We adopted the stochastic prioritization approach developed in [50] for experience replay. To address the exploration vs. exploitation dilemma, we deployed epsilon-greedy action selection, where the epsilon value was 0.3 at initiation and gradually decreased down to 0.1. Our framework was developed and trained on a NVIDIA GeForce RTX 2080/PCIe/SSE2 with dedicated 8 GB memory using an Intel® Core™ i7-9700K CPU @ 3.60 GHz × 8 processor along with 16 GB RAM. The development and training of the models was performed in PyTorch version 1.7.1 using CUDA version 11.4, cuDNN version 8.2.2, and NVIDIA Driver version 470.103.01.

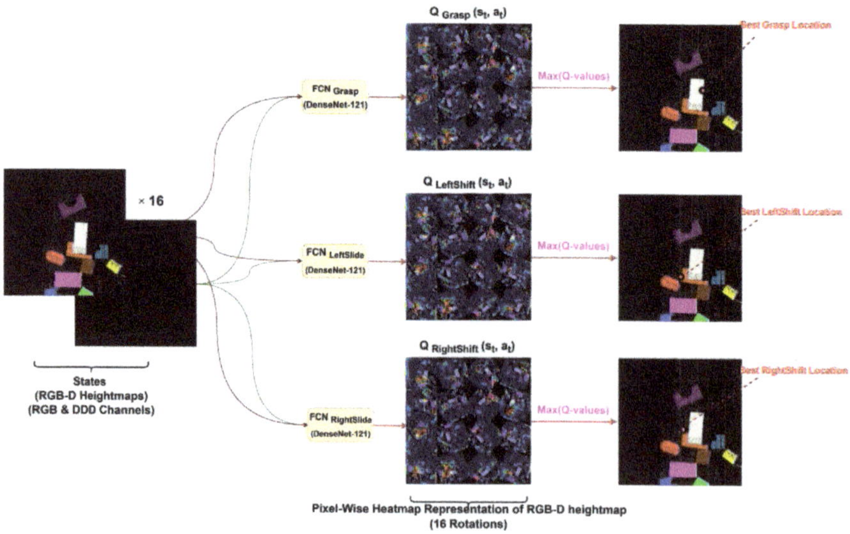

Figure 8. The flow of the whole approach is shown in this figure. RGB-D heightmaps comprising RGB and DDD channels generated as state representations are rotated by 22.5 degrees and all 16 rotations are fed to all three FCNs $\{FCN_{Grasp}, FCN_{LeftSlide}, FCN_{RightSlide}\}$ based on DenseNet-121 architecture. A pixel-wise map of affordances is generated by each FCN against each rotation of RGB-D heightmap. This map constitutes Q-values, shown in the heatmap format. The maximum Q-value is chosen out of all 48 pixel-wise maps for relevant action to be performed.

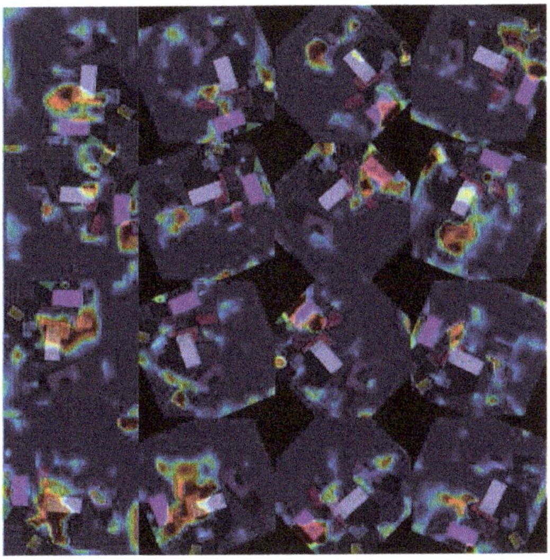

Figure 9. Heatmap visualizations of pixel-wise maps of Q-values generated against 16 rotations of an RGB-D heightmap at 22.5 degrees.

DenseNet is one of the most efficient architectures due to its maximized links leading to maximum feature reuse. However, at the same time, if it is not managed properly, due to its contagious convolutional operations its features, maps can start growing quadratically, thus creating a bottleneck in the GPU. Therefore, we implemented a memory-efficient variant as in [19], where shared memory schemes are used for concatenation, batch normalization, and gradients.

In our simulated environment, all agents are trained via self-supervision through trial and error. At the start of the training process, the workspace on the conveyer belt has N number of regular and irregular objects in a random clutter. The agent starts trying to pick and place these objects and keeps on doing so until the workspace is cleared of all the objects. Once all objects have been placed inside the basket, one iteration is completed and a new random clutter of objects appears in the workspace for the next iteration, and the process goes on. In our training, we kept the value of N = 10 for training, which means a random clutter of 10 different regular and irregular 3D objects (shown in Figure 10) in each iteration.

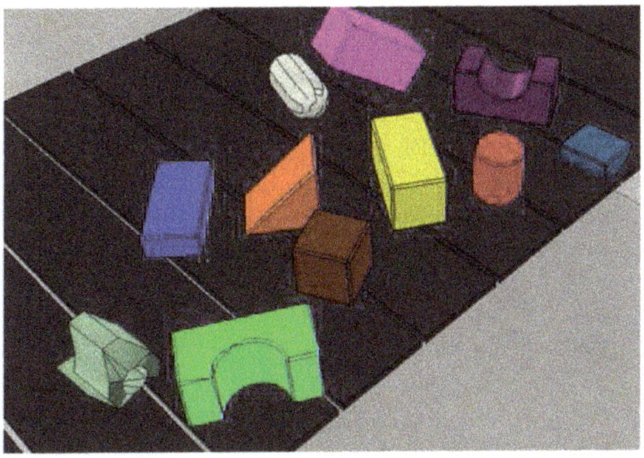

Figure 10. Ten regular and irregular 3D objects randomly used in the approach.

3.3. Simulation and Motion Planning Specifications

The Virtual Robot Experimentation Platform (V-REP) is a 3D robotic simulator with integrated development and coding support [51]. It also has the physics engines Bullet and ODE, providing a real-time experience of the objects involved in the simulation [52,53]. The API and threaded/non-threaded Lua scripting functionalities make it a reasonable option for combining multiple platforms such as python, java, C++, etc., for experimentation. Using this API, we were able to make our python-based RL agent communicate with the Lua-scripted simulated environment.

V-REP provides various motion planning calculation modules including the forward and inverse kinematics module. Forward kinematics means using kinematic equations, taking joints parameters as inputs, to calculate the position of an end-effector, whereas inverse kinematics is the reverse process, calculating joint parameters for the given position of an end-effector [54]. The collision detection module is another important module in V-REP used to highlight collisions if there are any. We used the UR5 robotic arm [55], a well-known cobot, for our approach, which is a 6-degrees-of-freedom robotic arm, and we used an RG2 gripper for grasping and sliding purposes. The whole simulation scene designed in V-REP can be seen in Figure 11. A conveyor belt was used in our experimentation. A ray-type infrared sensor is installed at the workspace end of the conveyor belt. The conveyor belt keeps moving, and as soon as the clutter of objects reaches the infrared sensor and breaks the ray, they are detected by the sensor. The sensor resultantly sends a signal to the conveyor belt controller to stop and to the vision sensor controller to capture the workspace RGB-D images for processing. As the workspace clears, the infrared sensor ray is reinstated and the conveyor belt starts moving again, thus bringing the next clutter, and in this way, the cycle continues.

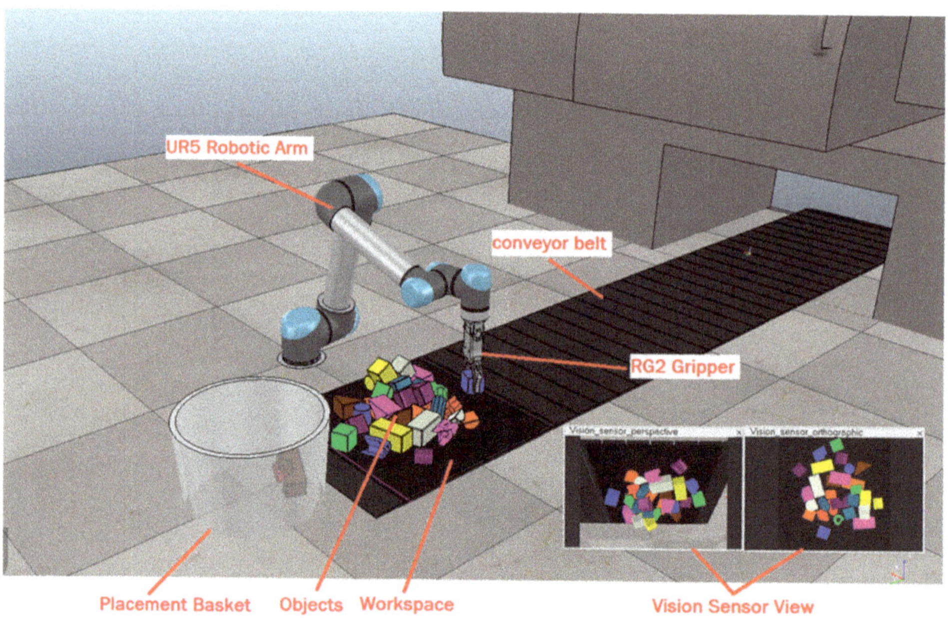

Figure 11. Whole simulation scene designed in V-REP.

We explored a number of options for performing our motion planning task such as OpenRave and Trajpot [56,57]. However, the best fit with our simulated environment was the Open Motion Planning Library (OMPL) [58] as it provided us with a certain degree of customization and control over the motion planning algorithms. The OMPL contains a number of geometric and control-based planners. Some of the many sampling-based planners available are Single-query Bi-direction Lazzy (SBL), Expansive Space Trees (EST), Rapidly exploring Random Trees (RRT), and the Probabilistic Roadmap Method (PRM), along with their many variants. In a similar way, many commonly used path planning state-spaces are readily available in an implemented form. The planner we used for motion planning and path calculations is a single-query planner, a bidirectional variant of Rapidly exploring Random Trees (RRT), known as RRT-Connect [59]. The core logic behind RRT-Connect is to develop two RRTs, one at the start point and the other at the end, and then, connect them. Due to this, the RRT-Connect planner tends to outperform the RRT planner.

4. Experimental Results and Discussion

In this section, we present a series of experiments carried out in order to see to what extent our goals can be achieved. Through these experiments, we assess the learning ability of the agent to perform pick-and-place tasks, the efficiency of prehensile manipulation, and the role of non-prehensile manipulation in doing so.

To evaluate the performance of our agent, we compared it with a deep learning baseline framework. We name our original approach G&S, which is the abbreviation for grasp and slide. The baseline approach is quite like our original approach. It is also designed around the idea of pixel-wise parameterization, but it is a deep learning-only-based supervised binary classification approach, i.e., there is no RL component. The same method of state representation and the same number of actions are formulated. As the number of actions is the same in both cases, the same number of FCNs are used. Like the concept of the original approach, each FCN will generate a dense pixel-wise map of affordances for the corresponding action. The only difference will be that the FCNs will be trained with binary classification (0 and 1) in a supervised manner and will generate a dense pixel-wise map of affordance values between 0 and 1. The labels for supervision will be generated on the basis of the success of grasping and sliding so that if grasping is successful, leading to placement of the object in the basket, it will be labeled as 1, and otherwise, as 0; moreover, if a change is detected in the workspace scene after left- or right-slide action is performed, it will be considered successful and labeled 1, and otherwise, as 0. Backpropagation was also performed in the same manner as in our G&S approach, as gradient passing was only performed through a particular FCN and pixel p, which gave the maximum corresponding value in the pixel-wise dense map for action selection, whereas 0 loss was backpropagated for all the remaining pixels. This supervised learning approach policy is considered greedy as it will go for the action with the maximum pixel-wise affordance value at any timestep. This baseline approach revolves around the concept of a pixel-wise affordance-based grasping approach designed in [12], but for an accurate comparison, we used similar DenseNet-121 [44] architecture and weights from the pretraining on ImageNet [45]. The comparison of the performance of our agents trained from scratch using the G&S approach and the baseline binary classification approach can be seen in Figure 12. In this comparison, both agents were trained for 3000 episodes, where each episode involved 10 randomly chosen regular- and irregular-shaped objects and ended when all the objects were placed, or at least ten actions consecutively failed to bring any change to the workspace. The blue line represents the success rate of our G&S approach, which is around 84%, and next to it, the green line represents the success rate of the binary classification approach, which is around 57%. This measure of success rate is estimated as the number of successfully placed objects in the episode over the total number of actions performed in the episode. A probable explanation for the reduced efficiency of our baseline binary classification approach may be its greedy approach. Instead of building and following strategies based on future perspectives as our G&S approach does through reward schemes, it focuses only on selecting actions based on the maximum pixel-wise affordance value at the current timestep.

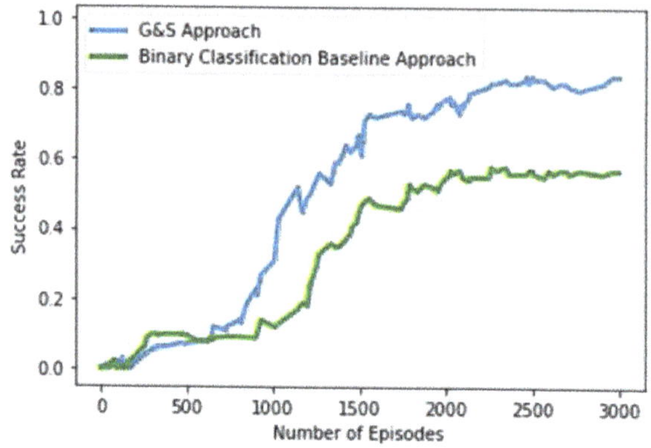

Figure 12. Performance comparison of original G&S approach and binary classification baseline approach.

We also designed another variant of our G&S approach, and the only difference between the original and this variant is that it uses the ResNet-101 architecture pre-trained on ImageNet [45] instead of DenseNet-121. ResNet [60], also known as a residual network, is made up of residual blocks. It was designed to tackle the fact that every network has a certain threshold value for its depth; as more and more layers pile up over the top of the network, its performance starts falling due to the notorious vanishing gradient problem. In the vanishing gradient problem, during backpropagation, as the gradient is transmitted through the depth, i.e., previous layers, it shrinks to an infinite level due to repetitive multiplications. ResNet resolves the vanishing gradient problem by introducing identity shortcut connections, also known as skip connections, which skip one or more layers, thus providing better gradient flow. The performance comparison of our original G&S approach based on DenseNet-121 pre-trained on ImageNet and its variant based on ResNet-101 pre-trained on ImageNet is shown in Figure 13. The original approach outperforms the ResNet-based variant by a margin of around 13%. The possible reason may be that ResNet performs a summation of feature maps of the layers, but DenseNet performs concatenation of feature maps, thus allowing for feature reusability at the maximum level. Whereas the identity shortcut connections restrict the representation capacity in ResNet, the multi-layer concatenation of feature maps through $L(L+1)/2$ links among the L layers in DenseNet enables it.

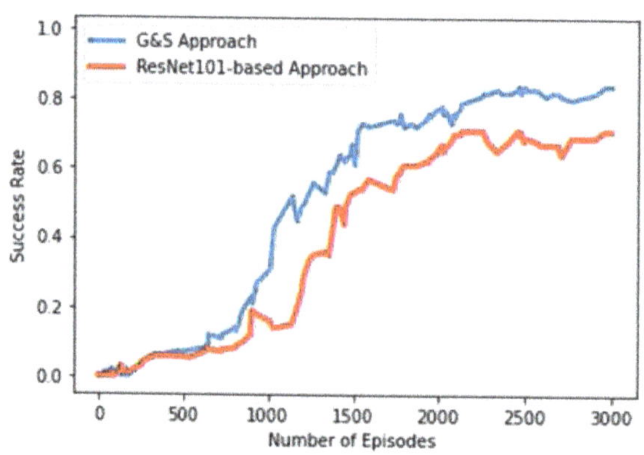

Figure 13. Performance comparison of original G&S approach and ResNet-based approach.

As the motive of our study is to train our agents in such a manner that they can learn the pick-and-place task through a combination of prehensile (grasping) and non-prehensile (left-slide and right-slide) robotic manipulations, we explored the role of non-prehensile actions in the convergence of the policy being learnt through trial and error under self-supervision. To understand the contribution of sliding manipulations to the overall learning of the pick-and-place task, we trained another variant of our original G&S approach with the difference that reward allocation for non-prehensile manipulations (left-slide and right-slide) was discontinued. Rewards were only awarded for prehensile manipulations (grasping) so that whenever an object was moved, the agent earned no reward. Figure 14 shows the performance of the original G&S approach in the blue line and the reward-less sliding variant in the cyan line. Removing rewards for sliding manipulations shows performance degradation of around 22% in terms of success rate. This highlights the importance of the contribution of non-prehensile manipulations in the learning of agents. The delayed rising curve of the success rate depicts the slower pace of the agent in learning the policy because the non-prehensile manipulation rewards are not there; thus, the agent is, by default, learning a policy that is more grasping-oriented. In this situation the importance of sliding manipulation can only be learnt indirectly through the rewards earned through grasping performed in the future after sliding manipulation; therefore, a lower rate of improvement is indicated by the delayed rising curve compared to the original G&S approach.

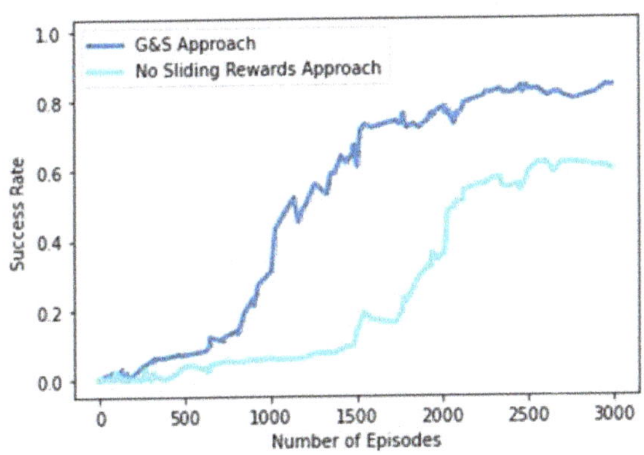

Figure 14. Performance comparison of original G&S approach and no-sliding-rewards approach.

In our G&S approach, we deployed DenseNet-121 architecture pre-trained on a popular ImageNet subset [61] comprising around 1,431,167 training, validation, and testing images in total belonging to around 1000 object classes. In our work, we also investigated the role and contribution of pre-trained weights, gained through training over ImageNet, in the learning process and the convergence of our learning policies by training a variant of our original G&S approach from scratch, in other words, involving no pre-trained weights. Figure 15 shows the success rates of both our original G&S approach in the blue line and its variant without pre-trained weights in an orange color. The results show that these pre-trained weights do not play much of a part in the process of learning the pick-and-place task, as the performance of both agents' is similar. Usually, pre-trained weights are involved in an architecture to decrease training time and increase accuracy. However, this scheme of using pre-trained weights does not make much difference in our case. We surmise that the likely reason may be the different kind of pixels pattern in our RGB-D heightmaps generated by following the approach presented in [12] compared to those in the images of the ImageNet dataset. In Figure 15, one can see the success rate's early curve increase in the orange line, representing the variant with no pre-trained weight. The delayed curve increase of the original G&S approach, represented by the blue line, can be seen in Figure 15. We suspect that the likely reason behind the delayed curve increase is the fact that the agent's initial training steps were consumed to escape the local optima problem due to pre-trained weights.

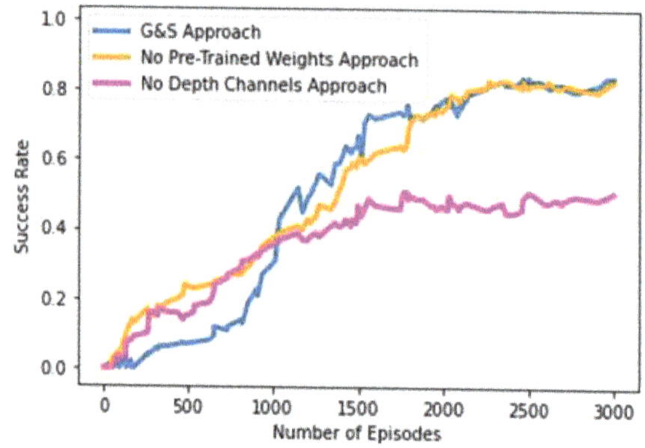

Figure 15. Performance comparison of original G&S approach, no-pre-trained-weights approach, and no-depth-channels approach.

Another element in our approach that is worth exploring is the depth information or the depth channels in our heightmaps. In Section 3.1, we explained that we cloned the depth across three channels to exploit weights pre-trained on ImageNet images. As such, as we already created a variant of our approach which is trained from scratch instead of using pre-trained weights, we removed the depth channels providing information regarding height from the bottom; then, for each FCN, we created only one trunk of DenseNet-121 to learn features from RGB channels instead of two trunks, because no second trunk was required to learn features from the cloned depth channels (DDD), and then, we further concatenated them with the features learnt from the RGB channels. In Figure 15, the magenta line represents the success rate of the variant of our G&S approach, which is trained from scratch without being provided with depth channels for feature extraction, and thus, no information regarding the height from the bottom. Excluding the feature learning from the depth channels led to performance deterioration as the success rate dropped to around 51% over the period of 3000 episodes. This clearly shows that the depth information or height-from-the-bottom information is quite crucial in the learning process of the agent and optimal convergence of the learning policies.

Further, we also reviewed our G&S approach's performance under unseen circumstances, which were not experienced during the training phases. As described in Section 3.2 during our training phases, the workspace on the conveyer belt has N number of regular and irregular objects in a random clutter and the agent starts picking and placing those objects and keeps on doing it until the workspace is cleared of all the objects. During the training phases, the number of random objects N was kept at 10. After training our agent for 3000 episodes consecutively on a clutter of 10 objects, we tested our agent by varying the clutter density. To bring variations in clutter density, we created three categories of clutter, as shown in Figure 16.

The first category is named the minimum density clutter, where regular and irregular objects are randomly distributed, and the number of objects ranges from 6 to 10. The second category is the medium density clutter, where the number of objects remains between 15 to 20 and they are distributed randomly. The third and last category, named the maximum density clutter, also comprises randomly distributed regular and irregular objects ranging from 25 to 30 in number. In addition to this, we designed several complicated cases comprising at least three and at maximum three objects, as shown in Figure 17, in order to test the behavior of our agent in these complicated scenarios. In these complicated scenarios, one or two objects were always kept free, whereas the remaining five to six regular and irregular objects were placed in some kind of locking manner so that these objects caused resistance while the agent tried to grasp or slide them.

Figure 16. Three categories based on clutter densities.

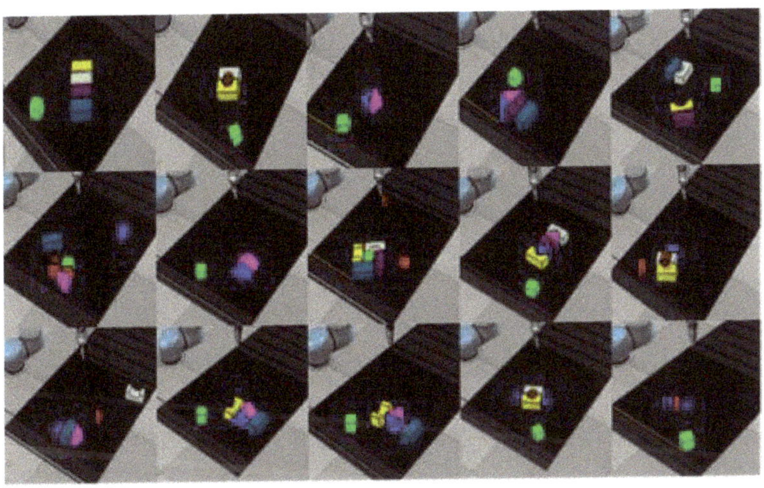

Figure 17. Complicated scenarios comprising 3 to 7 regular and irregular objects, designed for testing.

The average performance results of these clutter variations and complicated test cases are shown in Table 1.

Table 1. Average Test Results (%).

Category	Success Rate	Grasping Success
Minimum Clutter	84%	96%
Medium Clutter	82%	95%
Maximum Clutter	74%	82%
Complicated Scenarios	65%	73%

We can see in Table 1 that we compared the grasping success of our agent along with the overall success rate. This grasping success is calculated as the number of successful grasps over the total number of grasps performed in the whole episode. For instance, if 10 grasping actions were performed during the episode and 8 of them were successful, then grasping success will be calculated as $8/10 \times 100 = 80\%$. The results show that our agent performs well in minimum and medium clutter density test cases, but in maximum clutter density, performance degrades a little in both overall success rate and grasping success. The major reason behind this degradation of performance is the factor of objects appearing in the workspace on the conveyer belt at random locations, and due to high density, these objects are placed on each other in random weak poses. This is also visible in maximum density clutter, as shown in Figure 16. As soon as the robotic gripper goes to grasp an object that is weakly posed on other objects, they all slip and fall, leading to failed grasping and degradation of the overall performance. In the case of complicated scenarios, we intentionally placed objects in locking poses such that it became difficult for the agent to pick and place them. For instance, in some places, there was not enough space for grasping or sliding, and at the same time, some objects were too wide to be grasped. These scenarios were met by the agent for the very first time, as the object arrangements were completely random and not hand designed. Another reason for the decrease in overall performance is the smaller number of objects in the scenario. For instance, if the test case has only three objects, the agent clears the workspace in four actions, which is a very good performance, but still, the overall success rate will fall to $3/4 \times 100 = 75\%$. Therefore, we can say that the agent has truly shown its learning in our test cases in a very reasonable manner.

5. Conclusions and Future Work

In this study, we have presented a framework designed to enable intelligent and self-supervised industrial pick-and-place operations involving cobots. Our framework deals with the learning of both prehensile (grasping) and non-prehensile (sliding) robotic manipulations in a memory-efficient manner, therefore allowing the agent to learn and perform a sequential combination of prehensile and non-prehensile robotic manipulations to increase the efficiency and throughput of the pick-and-place task at hand. This approach deals with three unique actions and does not require any kind of prior domain knowledge or geometrical pose information like some previous approaches. In this our approach, we made our agent learn one prehensile manipulation—grasping—and two non-prehensile robotic manipulations—left-slide and right-slide. We used the pixel-wise parameterization method along with deep Q-learning to obtain pixel-wise maps of Q-values. The action with the highest Q-value is performed at the 3D location of the pixel that is predicted to have the maximum Q-value. Rewards are awarded accordingly for successful actions. Through the process of self-supervised trial and error, our agent learns and converges to the optimal policy, which enables the agent to perform successful non-prehensile manipulations at such moments where they lead to the possibility of successful prehensile manipulation in the future. Our agents learn these policies and perform successfully, as shown in our Section 4, where we show comparison results of different variants and baseline approaches and the test results for varying clutter densities and different test cases.

Thus, our approach shows many promising results, but at the same time, it also has some limitations. In our approach, the agent was trained on a limited number of 3D block shapes, which could be enhanced and extended to objects found in daily life such as bottles, cups, balls, etc. By doing so and redesigning the whole approach structure on ROS/ROS2, we intend to perform testing on a real UR5 robotic arm using daily life objects, too. Efficient management requirements due to the feature map concatenation factor of DenseNet-121 can also be seen as a kind of limitation. Another limitation can be seen in the factor of our approach only involving a sequential combination of robotic manipulations, for which, in the future, the number of actions can be increased and parallel robotic manipulation combinations can be brought in, along with other new robotic manipulations such as stacking, rolling, rotating, etc. By implementing these extensions, the system may become so large that the overestimation of future rewards will undoubtedly cause problems. Therefore, to resolve this issue, Double Q-learning [62] and Dueling Q-learning [63] variants can be explored in the future in order to increase the chances of efficiency and throughput.

Author Contributions: M.B.I. conceived and implemented the idea and wrote the paper; Y.Q. and B.L. reviewed and supervised the work. All authors have read and agreed to the published version of the manuscript.

Funding: This publication emanated from research conducted with the financial support of Science Foundation Ireland (SFI) under Grant Number SFI 16/RC/3918, co-funded by the European Regional Development Fund.

Institutional Review Board Statement: Not applicable.

Informed Consent Statement: Not applicable.

Data Availability Statement: Data are available on request from the authors.

Conflicts of Interest: The authors declare no conflict of interest.

References

1. Lamiraux, F.; Mirabel, J. Prehensile Manipulation Planning: Modeling, Algorithms and Implementation. *IEEE Trans. Robot.* **2021**, *38*, 2370–2388. [CrossRef]
2. Dogar, M.R.; Srinivasa, S.S. A Planning Framework for Non-Prehensile Manipulation under Clutter and Uncertainty. *Auton Robot.* **2012**, *33*, 217–236. [CrossRef]
3. Serra, D. Robot Control for Nonprehensile Dynamic Manipulation Tasks. Available online: https://www.researchgate.net/publication/310751102_Robot_Control_for_Nonprehensile_Dynamic_Manipulation_Tasks (accessed on 2 December 2022).
4. Weisz, J.; Allen, P.K. Pose error robust grasping from contact wrench space metrics. In Proceedings of the 2012 IEEE International Conference on Robotics and Automation, St Paul, MN, USA, 14–18 May 2012; pp. 557–562.
5. Pinto, L.; Gupta, A. Learning to Push by Grasping: Using multiple tasks for effective learning. *arXiv* **2016**, arXiv:1609.09025.
6. Hogan, F.R.; Rodriguez, A. Feedback Control of the Pusher-Slider System: A Story of Hybrid and Underactuated Contact Dynamics. *arXiv* **2016**, arXiv:1611.08268.
7. Mason, M.T. Mechanics and Planning of Manipulator Pushing Operations. *Int. J. Robot. Res.* **1986**, *5*, 53–71. [CrossRef]
8. Goyal, S.; Ruina, A.; Papadopoulos, J. Planar sliding with dry friction Part 1. Limit surface and moment function. *Wear* **1991**, *143*, 307–330. [CrossRef]
9. Clavera, I.; Held, D.; Abbeel, P. Policy Transfer via Modularity. Available online: https://davheld.github.io/DavidHeld_files/IROS___RL_pushing.pdf (accessed on 2 December 2022).
10. Bauzá, M.; Rodriguez, A. A probabilistic data-driven model for planar pushing. In Proceedings of the 2017 IEEE International Conference on Robotics and Automation (ICRA), Singapore, 29 May–3 June 2017.
11. Finn, C.; Levine, S. Deep visual foresight for planning robot motion. In Proceedings of the 2017 IEEE International Conference on Robotics and Automation (ICRA), Marina Bay Sands Convention Centre, Singapore, 29 May–3 June 2017; pp. 2786–2793.
12. Zeng, A.; Song, S.; Yu, K.-T.; Donlon, E.; Hogan, F.R.; Bauza, M.; Ma, D.; Taylor, O.; Liu, M.; Romo, E.; et al. Robotic pick-and-place of novel objects in clutter with multi-affordance grasping and cross-domain image matching. *Int. J. Robot. Res.* **2022**, *41*, 690–705. [CrossRef]
13. Pinto, L.; Gupta, A. Supersizing Self-supervision: Learning to Grasp from 50K Tries and 700 Robot Hours. In Proceedings of the 2016 IEEE International Conference on Robotics and Automation, Stockholm, Sweden, 16–21 May 2016.
14. Gualtieri, M.; ten Pas, A.; Platt, R.W. Category Level Pick and Place Using Deep Reinforcement Learning. *arXiv* **2017**, arXiv:1707.05615.
15. Mahler, J.; Liang, J.; Niyaz, S.; Laskey, M.; Doan, R.; Liu, X.; Ojea, J.A.; Goldberg, K. Dex-Net 2.0: Deep Learning to Plan Robust Grasps with Synthetic Point Clouds and Analytic Grasp Metrics. *arXiv* **2017**, arXiv:1703.09312.
16. Gualtieri, M.; ten Pas, A.; Saenko, K.; Platt, R. High precision grasp pose detection in dense clutter. *arXiv* **2017**, arXiv:1603.01564.
17. Lenz, I.; Lee, H.; Saxena, A. Deep Learning for Detecting Robotic Grasps. *Int. J. Robot. Res.* **2015**, *34*, 705–724. [CrossRef]
18. Boularias, A.; Bagnell, J.A.; Stentz, A. Learning to Manipulate Unknown Objects in Clutter by Reinforcement. In Proceedings of the Twenty-Ninth AAAI Conference on Artificial Intelligence, Austin, TX, USA, 25–30 January 2015; pp. 1336–1342.
19. Pleiss, G.; Chen, D.; Huang, G.; Li, T.; van der Maaten, L.; Weinberger, K.Q. Memory-Efficient Implementation of DenseNets. *arXiv* **2017**, arXiv:1707.06990.
20. Vaudel, C. Dobot V3 Conveyor Belt Kit. Available online: https://www.robotlab.com/store/conveyor-belt-kit-for-robotic-arm (accessed on 12 January 2023).
21. Conveyor Belt (Compatible with Ned2, Ned and Niryo One). Available online: https://www.generationrobots.com/en/403942-conveyor-belt-compatible-with-ned2-ned-and-niryo-one.html (accessed on 12 January 2023).
22. Sandra Bin Picking in the Industry. *ATRIA Innovation*. 2021. Available online: https://www.atriainnovation.com/en/bin-picking-in-the-industry/ (accessed on 2 December 2022).
23. Great Examples of Automated Bin Picking by Pick-it 3D—Pick-it 3D-Robot Vision Made Easy. Available online: https://www.pickit3d.com/en/blog/great-examples-of-pick-it-3d-bin-picking (accessed on 12 January 2023).

24. Eppner, C.; Höfer, S.; Jonschkowski, R.; Martín-Martín, R.; Sieverling, A.; Wall, V.; Brock, O. Lessons from the Amazon Picking Challenge: Four Aspects of Building Robotic Systems. In Proceedings of the 26th International Joint Conference on Artificial Intelligence, Melbourne, VIC, Australia, 19–25 August 2017; pp. 4831–4835.
25. Rodriguez, A.; Mason, M.T.; Ferry, S. From caging to grasping. *Int. J. Robot. Res.* **2012**, *31*, 886–900. [CrossRef]
26. Singh, T.; Ambike, S. A soft-contact and wrench based approach to study grasp planning and execution. *J. Biomech.* **2015**, *48*, 3961–3967. [CrossRef]
27. Zhou, J.; Liu, Y.; Liu, J.; Xie, Q.; Zhang, Y.; Zhu, X.; Ding, X. BOLD3D: A 3D BOLD descriptor for 6Dof pose estimation. *Comput. Graph.* **2020**, *89*, 94–104. [CrossRef]
28. Goldfeder, C.; Ciocarlie, M.; Dang, H.; Allen, P.K. The Columbia grasp database. In Proceedings of the 2009 IEEE International Conference on Robotics and Automation, Kobe, Japan, 12–17 May 2009; pp. 1710–1716.
29. Redmon, J.; Angelova, A. Real-Time Grasp Detection Using Convolutional Neural Networks. In Proceedings of the 2015 IEEE International Conference on Robotics and Automation, Seattle, WA, USA, 26–30 May 2015.
30. Stüber, J.; Zito, C.; Stolkin, R. Let's Push Things Forward: A Survey on Robot Pushing. *Front. Robot. AI* **2020**, *7*, 8. [CrossRef]
31. Zhou, J.; Mason, M.T.; Paolini, R.; Bagnell, D. A convex polynomial model for planar sliding mechanics: Theory, application, and experimental validation. *Int. J. Robot. Res.* **2018**, *37*, 249–265. [CrossRef]
32. Omrcen, D.; Boge, C.; Asfour, T.; Ude, A.; Dillmann, R. Autonomous acquisition of pushing actions to support object grasping with a humanoid robot. In Proceedings of the 2009 9th IEEE-RAS International Conference on Humanoid Robots, Paris, France, 7–10 December 2009; pp. 277–283.
33. Danielczuk, M.; Mahler, J.; Correa, C.; Goldberg, K. Linear Push Policies to Increase Grasp Access for Robot Bin Picking. In Proceedings of the 2018 IEEE 14th International Conference on Automation Science and Engineering (CASE), Munich, Germany, 20–24 August 2018; pp. 1249–1256.
34. Sarantopoulos, I.; Kiatos, M.; Doulgeri, Z.; Malassiotis, S. Total Singulation with Modular Reinforcement Learning. *IEEE Robot. Autom. Lett.* **2021**, *6*, 4117–4124. [CrossRef]
35. Kiatos, M.; Malassiotis, S. Robust object grasping in clutter via singulation. In Proceedings of the 2019 International Conference on Robotics and Automation (ICRA), Montreal, QC, Canada, 20–24 May 2019; pp. 1596–1600.
36. Imtiaz, M.B.; Qiao, Y.; Lee, B. Prehensile Robotic pick-and-place in clutter with Deep Reinforcement Learning. In Proceedings of the 2022 International Conference on Electrical, Computer and Energy Technologies (ICECET), Cape Town, South Africa, 9–10 December 2022; IEEE: Prague, Czech Republic; pp. 1–6.
37. Mohammed, M.Q.; Chung, K.L.; Chyi, C.S. Pick and Place Objects in a Cluttered Scene Using Deep Reinforcement Learning. *Int. J. Mech. Mechatron. Eng.* **2020**, *20*, 50–57.
38. Zeng, A.; Song, S.; Welker, S.; Lee, J.; Rodriguez, A.; Funkhouser, T.A. Learning Synergies between Pushing and Grasping with Self-supervised Deep Reinforcement Learning. In Proceedings of the 2018 IEEE/RSJ International Conference on Intelligent Robots and Systems (IROS), Madrid, Spain, 1–5 October 2018; pp. 4238–4245.
39. Cruz, L.; Lucio, D.; Velho, L. Kinect and RGBD Images: Challenges and Applications. In Proceedings of the 2012 25th SIBGRAPI Conference on Graphics, Patterns and Images Tutorials, Ouro Preto, Brazil, 22–25 August 2012; pp. 36–49.
40. UR+ | RG2 Flexible Collaborative Gripper. Available online: https://www.universal-robots.com/plus/products/onrobot/rg2-gripper/ (accessed on 17 February 2022).
41. Mnih, V.; Kavukcuoglu, K.; Silver, D.; Graves, A.; Antonoglou, I.; Wierstra, D.; Riedmiller, M.A. Playing atari with deep reinforcement Learning. *arXiv* **2013**, arXiv:1312.5602.
42. Mnih, V.; Kavukcuoglu, K.; Silver, D.; Rusu, A.A.; Veness, J.; Bellemare, M.G.; Graves, A.; Riedmiller, M.; Fidjeland, A.K.; Ostrovski, G.; et al. Human-level control through deep reinforcement learning. *Nature* **2015**, *518*, 529–533. [CrossRef]
43. Long, J.; Shelhamer, E.; Darrell, T. Fully Convolutional Networks for Semantic Segmentation. In Proceedings of the IEEE Conference on Computer Vision and Pattern Recognition, Columbus, OH, USA, 23–28 June 2014.
44. Huang, G.; Liu, Z.; Weinberger, K.Q. Densely connected convolutional networks. In Proceedings of the IEEE Conference on Computer Vision and Pattern Recognition, Las Vegas, NV, USA, 27–30 June 2016.
45. Deng, J.; Dong, W.; Socher, R.; Li, L.-J.; Li, K.; Fei-Fei, L. ImageNet: A large-scale hierarchical image database. In Proceedings of the 2009 IEEE Conference on Computer Vision and Pattern Recognition, Miami, FL, USA, 20–25 June 2009; pp. 248–255.
46. Nair, V.; Hinton, G.E. Rectified Linear Units Improve Restricted Boltzmann Machines. In Proceedings of the Proceedings of the 27th International Conference on International Conference on Machine Learning, Madison, WI, USA, 21–24 June 2010; pp. 807–814.
47. Ioffe, S.; Szegedy, C. Batch Normalization: Accelerating Deep Network Training by Reducing Internal Covariate Shift. In Proceedings of the 32nd International Conference on International Conference on Machine Learning, Lille, France, 6–11 July 2015; Volume 37, pp. 448–456.
48. Huber, P.J. Robust Estimation of a Location Parameter. *Ann. Math. Stat.* **1964**, *35*, 73–101. [CrossRef]
49. Rumelhart, D.E.; Hintont, G.E.; Williams, R.J. Learning representations by back-propagating errors. *Nature* **1986**, *323*, 533–536. [CrossRef]

50. Schaul, T.; Quan, J.; Antonoglou, I.; Silver, D. Prioritized experience replay. *arXiv* **2015**, arXiv:1511.05952.
51. Freese, M.; Singh, S.; Ozaki, F.; Matsuhira, N. Virtual Robot Experimentation Platform V-REP: A Versatile 3D Robot Simulator. In *Simulation, Modeling, and Programming for Autonomous Robots*; Ando, N., Balakirsky, S., Hemker, T., Reggiani, M., von Stryk, O., Eds.; Lecture Notes in Computer Science; Springer: Berlin/Heidelberg, Germany, 2010; Volume 6472, pp. 51–62. ISBN 978-3-642-17318-9.
52. Imtiaz, M.; Qiao, Y.; Lee, B. Comparison of Two Reinforcement Learning Algorithms for Robotic Pick and Place with Non-Visual Sensing. *Int. J. Mech. Eng. Robot. Res.* **2021**, *10*, 526–535. [CrossRef]
53. Imtiaz, M.B.; Qiao, Y.; Lee, B. Implementing Robotic Pick and Place with Non-visual Sensing Using Reinforcement Learning. In Proceedings of the 2022 6th International Conference on Robotics, Control and Automation (ICRCA), Xiamen, China, 12–14 January 2022; pp. 23–28.
54. Zou, Z.; Han, J.; Zhou, M. Research on the inverse kinematics solution of robot arm for watermelon picking. In Proceedings of the 2017 IEEE 2nd Information Technology, Networking, Electronic and Automation Control Conference (ITNEC), Chengdu, China, 15–17 December 2017; pp. 1399–1402.
55. SC15F, N.; Nachi, A.; Controller, L.-L.; NACHI, N.; Controller, H.-L. 2.4 Universal Robots UR5. *Sens. Based Real-Time Control Ind. Robot.* **2013**, *16*. Available online: https://ntnuopen.ntnu.no/ntnu-xmlui/handle/11250/261111 (accessed on 2 December 2022).
56. Diankov, R. Automated Construction of Robotic Manipulation Programs. Ph.D. Thesis, Carnegie Mellon University, Pittsburgh, PA, USA, 2010; 263p.
57. Schulman, J.; Ho, J.; Lee, A.; Awwal, I.; Bradlow, H.; Abbeel, P. Finding Locally Optimal, Collision-Free Trajectories with Sequential Convex Optimization. In Proceedings of the Robotics: Science and Systems IX, Robotics: Science and Systems Foundation, Berlin, Germany, 24–28 June 2013.
58. Sucan, I.A.; Moll, M.; Kavraki, L.E. The Open Motion Planning Library. *IEEE Robot. Automat. Mag.* **2012**, *19*, 72–82. [CrossRef]
59. Kuffner, J.J.; LaValle, S.M. RRT-connect: An efficient approach to single-query path planning. In Proceedings of the Proceedings 2000 ICRA. Millennium Conference. IEEE International Conference on Robotics and Automation, San Francisco, CA, USA, 24–28 April 2000; Symposia Proceedings (Cat. No.00CH37065). Volume 2, pp. 995–1001.
60. He, K.; Zhang, X.; Ren, S.; Sun, J. Deep Residual Learning for Image Recognition. *arXiv* **2015**, arXiv:1512.03385.
61. Krizhevsky, A.; Sutskever, I.; Hinton, G.E. ImageNet Classification with Deep Convolutional Neural Networks. In Proceedings of the Proceedings of the 25th International Conference on Neural Information Processing Systems, Lake Tahoe, NV, USA, 3–6 December 2012; Curran Associates Inc.: Red Hook, NY, USA, 2012; Volume 1, pp. 1097–1105.
62. van Hasselt, H.; Guez, A.; Silver, D. Deep Reinforcement Learning with Double Q-Learning. In Proceedings of the Thirtieth AAAI Conference on Artificial Intelligence, Phoenix, AZ, USA, 12–17 February 2016; AAAI Press: Palo Alto, CA, USA, 2016; pp. 2094–2100.
63. Wang, Z.; de Freitas, N.; Lanctot, M. Dueling Network Architectures for Deep Reinforcement Learning. In Proceedings of the International Conference on Machine Learning, New York, NY, USA, 19–24 June 2016.

sensors

MDPI

Article

Research on Surface Tracking and Constant Force Control of a Grinding Robot

Xiaohua Shi [1,*], Mingyang Li [2], Yuehu Dong [1] and Shangyu Feng [1,3]

[1] School of Mechanical Engineering, Yanshan University, Qinhuangdao 066000, China; qqxr1998@126.com (Y.D.); fengshangyu@inovance.com (S.F.)
[2] School of Vehicle and Energy, Yanshan University, Qinhuangdao 066000, China; limingyang1998@126.com
[3] Suzhou Huichuan Control Technology Co., Ltd., Suzhou 215000, China
* Correspondence: xhshi@ysu.edu.cn; Tel.: +86-150-3353-5740

Abstract: To improve the quality and efficiency of robot grinding, a design and a control algorithm for a robot used for grinding the surfaces of large, curved workpieces with unknown parameters, such as wind turbine blades, are proposed herein. Firstly, the structure and motion mode of the grinding robot are determined. Secondly, in order to solve the problem of complexity and poor adaptability of the algorithm in the grinding process, a force/position hybrid control strategy based on fuzzy PID is proposed which greatly improves the response speed and reduces the error of the static control strategy. Compared with normal PID, fuzzy PID has the advantages of variable parameters and strong adaptability; the hydraulic cylinder used to adjust the angle of the manipulator can control the speed offset within 0.27 rad/s, and the grinding process can be carried out directly without obtaining the specific model of the surface to be machined. Finally, the experiments are carried out, the grinding force and feed speed are maintained within the allowable error range of the expected value, and the results verify the feasibility and effectiveness of the position tracking and constant force control strategy in this paper. The surface roughness of the blade is maintained within Ra = 2~3 μm after grinding, which proves that the grinding quality meets the requirements of the best surface roughness required for the subsequent process.

Keywords: grinding robot; constant force control; surface tracking; fuzzy PID control; wind turbine blade

Citation: Shi, X.; Li, M.; Dong, Y.; Feng, S. Research on Surface Tracking and Constant Force Control of a Grinding Robot. *Sensors* **2023**, *23*, 4702. https://doi.org/10.3390/s23104702

Academic Editors: Xin Zhao, Mingzhu Sun and Qili Zhao

Received: 17 April 2023
Revised: 5 May 2023
Accepted: 8 May 2023
Published: 12 May 2023

1. Introduction

The development of the wind power industry is of great significance to environmental protection, and there has been rapid development, with increasing requirements for wind turbines in recent years [1]. Therefore, the manufacturing process of wind turbine blades and other components is gradually becoming more intelligent and automated [2], and the process of grinding wind turbine blades is essential. The roughness after grinding affects the blade's subsequent effectiveness, and the grinding process itself determines the production cost and service life of the blade. Due to the irregular shape and large size of a wind turbine blade's surface, artificial grinding, which is currently the most common method of treatment, is not merely inefficient, but also struggles to achieve uniformity, and polished glass fiber dust can cause environmental pollution and damage to workers' health [3,4]. In consequence, the design of, and research into, wind turbine blade grinding robots has gradually attracted more interest in recent years. Some of them employ a gantry structure [5,6] equipped with a grinding head on each side, whereby the gantry moves along a guide rail to perform the grinding of the blades. This model achieves high grinding efficiency but requires much space. Others use mobile platforms equipped with multi-degree-of-freedom manipulators and grinding devices [7], which can effectively adapt to the site, but with more difficult manipulator control.

With the gradual increase in industrial automation in recent years, the requirements relating to the control and repetitive positioning accuracy of robots have increased, and the environmental aspects of robots need to be improved. The surfaces of wind turbine blades are complex, with curvature variation in the axial and circumferential directions; thus, the control of the grinding robot is particularly important. To ensure a suitable grinding effect, it is essential to maintain a constant grinding force and the real-time tracking of the blade surface.

Most previous studies have focused on force control, which can be divided into passive control and active control [8]. Guo et al. [9] proposed a new two-dimensional adjustable force mechanism that does not require additional sensors. However, because of the low accuracy and poor stability of this method of passive constant force control, active force control has become a research hotspot.

Active force control includes impedance control and force–position hybrid control. Shen et al. [10] proposed a fuzzy-based impedance control algorithm, with which the contact force error could be controlled within 2N for the external environment with unknown stiffness. Wahballa et al. [11] used an OSRDF impedance controller to offset the damping force and the contact force tracking error, which improved the system stability.

PID control is a common method applied to achieve active force control. The process is proportional, integral and differential, enabling quick responsiveness and high stability in control, and the robustness of the mode of PID control is strong. To resolve the issues of model construction and algorithm complexity that arise in the active flexibility control method, Yao et al. [12] proposed a fuzzy PID smooth grinding control algorithm based on a two-degree-of-freedom constant force grinding device driven by pneumatic force, wherein the position and force are decoupled and controlled separately, and verified that the control precision and response time of the fuzzy PID control algorithm were better than those of the conventional PID control method. In addition, Sun et al. [13] confirmed the effectiveness of the fuzzy PID control algorithm. Dai et al. [14] also presented a backstepping + PID control method to closely track the applied force.

As such, many scholars use PID as the control strategy in their research. Xu et al. [15] proposed the combination of an active strategy, based on PI/PD control, and a passive strategy, based on PID control, which was shown to improve the accuracy and efficiency of the controlled force and avoid over- and under-cutting. Zhang et al. [16] presented a method for optimizing the parameters of a robot PD constant grinding force controller using deep reinforcement learning DRL Rainbow, which can ensure that a constant force is applied during grinding by adjusting the grinding depth. Further, some studies have been carried out on the normal vector of the surface; in establishing the positioning of the grinding tool during processing, Zhao et al. [17] proposed an adaptive PD constant force controller and a normal vector search algorithm, which together ensure that constant force is applied during the grinding of workpieces with unknown shapes. Wang et al. [18] put forward a force–position hybrid control method based on a PD constant force control algorithm. Han et al. [19] presented a fuzzy gain scheduling PID controller, which could reduce the influence of dynamic characteristics on the control system.

As regards the influence of curvature variance, beginning with an improved material removal model, Li et al. [20] took the material to be removed from different positions on the object being machined as the research object, and constructed a hybrid force–position control algorithm to ensure the contour accuracy of workpieces with complex curved surfaces.

Some scholars have proposed other new algorithms to ensure constant force control. In order to resolve the shortcomings of traditional robots' constant force output, Jia et al. [21] combined the PID control method with a particle swarm optimization algorithm (PSO) to realize the constant force output of a cylinder and drive the grinding device. Zhao et al. [22] developed a strategy for constant force grinding that reconstructs the surface of the workpiece based on the point cloud data obtained via visual measurements, without requiring an accurate CAD model of the workpiece. In the grinding process, noise in the machining environment will affect the constant force control of the robot. Dai et al. [23] proposed a new force control method using an extended state observer (ESO) that can reduce fluctuations in the grinding force in the normal direction. Due to the impact between the grinding tool and the surface being machined, there is constant fluctuation in the contact force, and so Xiao et al. [24] put forward an active disturbance rejection method. Here, in order to achieve constant force control during machining, an RBF neural network model and an iterative algorithm are combined. Zhang [25] presented a variable-gain iterative learning force–position hybrid control method with ideal contour force as the contact point processing idea, which could still ensure the control accuracy of contour force when the tracking error was not accurate enough. Li et al. [26] used the force–position hybrid control method to estimate and adjust the contact state between the robot and the unknown workpiece, which could provide a constant normal contact force. Zhang et al. [27] proposed a constant force grinding controller based on proximal strategy optimization, which improved the response ability of the system.

The roughness of the outer surface of the wind turbine blade before grinding is about Ra = 0.8~1.5 μm, and the requirements are met when this value after grinding is Ra = 1.5~8 μm. A roughness of Ra = 2.0~3.0 μm yields the best results in the subsequent painting of the blade.

In this paper, a wind-turbine-blade-grinding robot, along with its control strategy and related algorithms, are introduced. Given the size and complexity of the surfaces of wind turbine blades, this paper considers the surface tracking and constant force control of blades simultaneously, so as to enhance the surface quality of wind turbine blades after grinding. Firstly, the structure of the grinding robot is described, and the method for planning the motion of the robot is designed. Then, the strategies of surface tracking and constant force control are established; fuzzy PID control is applied to the power source of the manipulator, and gravity compensation for the grinding robot is conducted. Finally, the controlling algorithm constructed in this paper is applied to the operation of a grinding robot to validate the reliability of our method of surface tracking and constant force control; by comparison with artificial grinding, the feasibility of the effectiveness of the scheme is confirmed. The experimental results are discussed at the end of the paper.

2. Grinding Robot

2.1. Structure of the Grinding Robot

The grinding robot addressed in this paper is composed of three main parts: an omnidirectional motion chassis, a five-degree-of-freedom manipulator and a flexible grinding device, as shown in Figure 1. The omnidirectional motion chassis is equipped with a five-degree-of-freedom manipulator to enable the maneuverability of the grinding robot, and a flexible grinding device is installed at the end of the manipulator.

Figure 1. Grinding robot.

The structure of the pose adjustment system of the flexible grinding device is shown in Figure 2. Four small springs are arranged around the directional connecting block in the pose adjustment system to ensure that the flexible grinding device can move around the axial surface of the blade, and a micro displacement sensor is included under the large spring in the center, so the normal force of the grinding device on the blade is governed by measuring the compression distance of this spring.

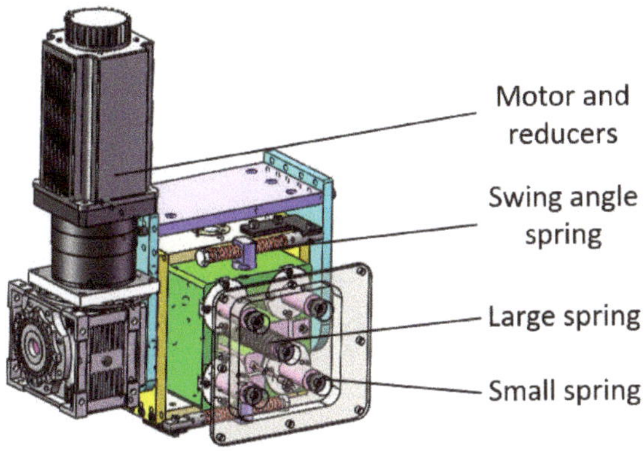

Figure 2. Pose adjustment system.

2.2. Motion Planning and Control Scheme

Unlike the structure of a traditional manipulator, the wind-turbine-blade-grinding robot addressed in this paper is a type of vehicle-mounted combined manipulator. For simplicity, the control of the total motion of the grinding robot is separated into two steps. First is the position adjustment of the whole machine; the grinding robot is moved to the correct area for blade processing in this stage. Here, only the control of the moving chassis is considered, while the control of the manipulator is not considered. Second is the grinding stage, during which the mechanical arm maneuvers the front-end grinding device to carry out the grinding of the blade. At this stage, the moving chassis is stationary, and only the control of the motion and force of the mechanical arm are considered, so as to ensure the surface tracking of, and the control of force applied to, the blade. After completing a given grinding area, the above stages are repeated in the next area. In this paper, a longitudinal-spacing grinding method is adopted, and the planning of the overall motion of the grinding robot is illustrated in Figure 3.

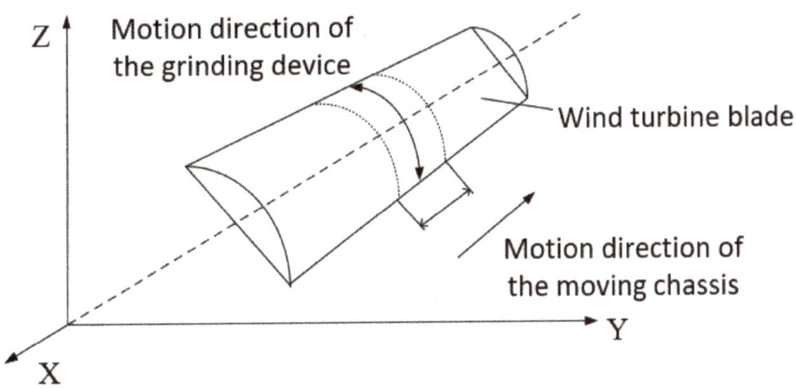

Figure 3. Motion planning of the grinding robot.

In the first stage, only the position of the moving chassis needs to be adjusted, and must be kept at a distance of 1.5~2.5 m from the blade. Figure 4 shows the control method employed in this stage.

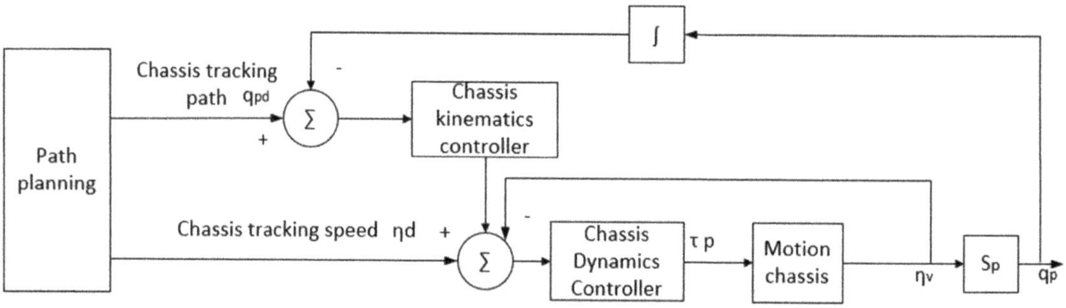

Figure 4. Strategy of controlling the grinding robot in the position-adjustment stage.

In the second stage of the grinding operation, due to the presence of different normal vectors at each grinding point on the outer surface of the wind turbine blade, each part of the grinding robot that can be manipulated needs to be coordinated and controlled to enable pose adjustment. Based on the Peterson equation, the empirical formula of the material removal rate during the grinding process [28,29] is

$$r = C_A K_A k_t \frac{V_b}{V_w l_w} F_A \tag{1}$$

where r is the instantaneous removal rate of material, C_A is a constant determined via experimentation, K_A is the resistance coefficient, k_t is the durability coefficient of the grinding tool, V_b is the cutting speed, V_w is the feed rate, l_w is the width of the grinding area and F_A is the grinding force. The factors that most determine the impact of the grinding robot on the blade are the grinding force and the feed rate, so these two factors should be controlled in all following steps. The control strategy applied is shown in Figure 5, and this strategy enables real-time surface tracking and the constant control of the grinding force.

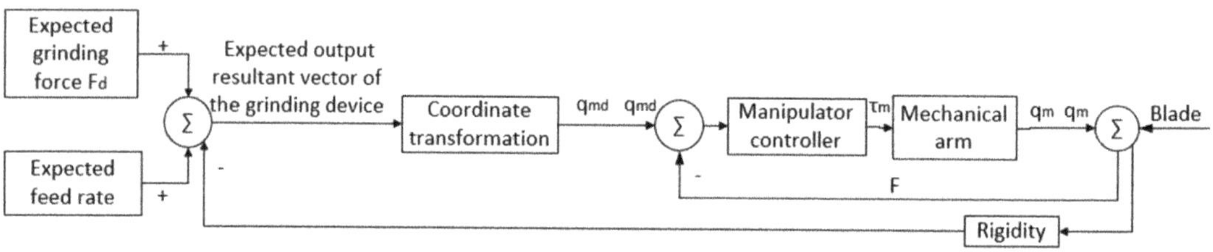

Figure 5. Control strategy of grinding robot applied in the grinding stage.

In summary, the control scheme applied to the grinding robot, in which the moving chassis, the manipulator and the front-end grinding device are hierarchically controlled, is shown in Figure 6. The commands controlling the chassis are sent via a remote controller, and each wheel group is controlled separately. Data on the distance and angle of movement are collected by the encoder and angle sensors and are sent back to the host computer. Information on the positioning of the manipulator and the grinding device is collected by multiple sensors and transmitted to the host computer. The two laser sensors installed on the upper and lower edges of the grinding device ensure that the tangent of the grinding roller and blade is perpendicular to the normal vector of the grinding point of the blade, and the displacement sensor included in the pose adjustment system described above controls the pressure of the grinding device on the blade. Cable sensor 1 on the lifting part of the manipulator, the angle sensor incorporated in the angle adjustment mechanism, and cable sensor 2 on the telescopic part jointly control the grinding force and speed of the robot.

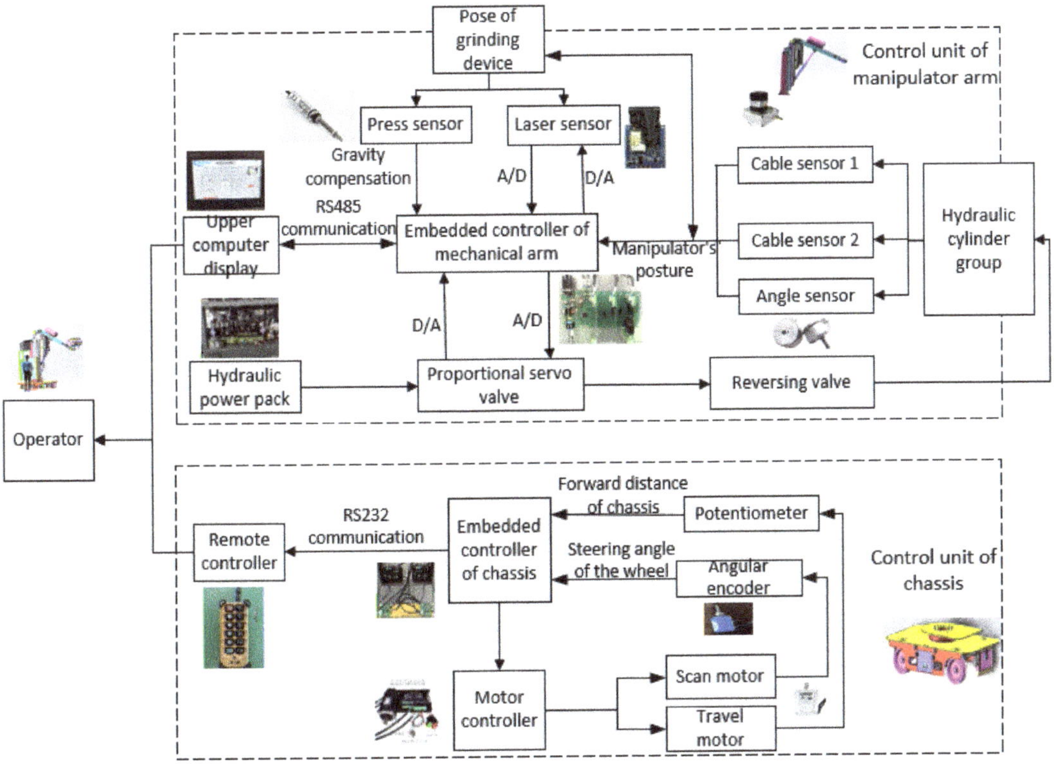

Figure 6. Scheme of control unit.

3. Control Strategy of Grinding Robot

In the above analysis, the grinding robot is divided into two parts: the manipulator and the moving chassis, the latter of which can be controlled by the human operator. Therefore, control of the chassis is not addressed in detail in the subsequent discussion, wherein we consider the real-time tracking of the blade's surface and the maintenance of consistent grinding via the control of the manipulator.

3.1. Surface Tracking Control Strategy

3.1.1. Surface Tracking of Blade Surface

Due to the changing curvature and irregular shape of wind turbine blades, it is very important for a grinding robot to track its position on a blade's surface during grinding, and so it is necessary to establish a reasonable control strategy [30]. Two PANFEE L2S laser ranging sensors are included above and below the front-end polishing device to track the blade surface, as shown in Figure 7a.

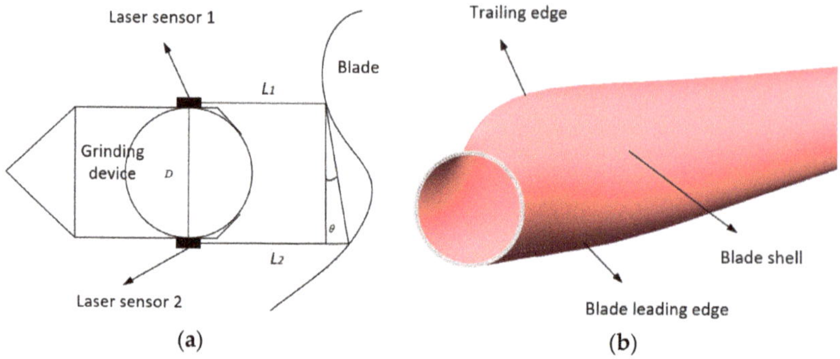

Figure 7. Surface tracking diagram. (**a**) Surface tracking method; (**b**) Different grinding areas of blade.

Here, the parameters L_1 and L_2 represent the distance between the grinding device and the surface of the wind turbine blade as measured by the laser sensors above and below the grinding device, respectively; D represents the distance between the two laser sensors, and θ represents the angle of offset between the attitude of the grinding device and the deviation of the blade, the calculation formula of which is

$$\theta = \arctan\left(\frac{L_2 - L_1}{D}\right) \tag{2}$$

After calculating θ, the pitch angle of the grinding device can be adjusted using fuzzy PID control until θ reaches zero and the parameters L_1 and L_2 are equal, which ensures that the grinding device is parallel to the tangent line of the blade's surface at the point of grinding, and this process is carried out in real time during grinding.

In addition, when the two relevant degrees of freedom—of the three degrees of freedom of lifting, stretching and angle adjustment in the manipulator—are determined, the end of the grinding roller can be moved along any trajectory. Therefore, in order to optimize the control algorithm, the position surface tracking of the robot is planned according to the three limit conditions in the grinding process, as shown in Figure 7b.

Firstly, when the grinding area is the trailing edge of the blade, the grinding surface is almost an inclined plane and the curvature change is comparatively small. Therefore, combining the two degrees of freedom of lifting and stretching, the robot can realize the grinding work at the trailing edge of the blade.

Secondly, when the grinding area is the middle shell, the curvature of the blade begins to change significantly, and the effect of the telescopic manipulator is small, so the degrees of freedom of lifting and angle adjustment are used at this time.

Finally, when the grinding area is the leading edge of the blade, the surface shape is convex, and it is also the bottom end of a single grinding process. At this time, the lifting degree of freedom no longer applies; in consequence, combining the two degrees of freedom of stretching and angle adjustment, the grinding action can satisfy the surface of the leading edge of the blade.

In this way, the tracking of blade surface in a single grinding process can be realized.

3.1.2. Adjustment of the Robot's Position

To ensure the safe operation of the grinding robot and improve the overall performance, the grinding robot must keep a safe vertical distance from the blade (set at about 1.5~2.5 m in this paper), which is measured using laser sensor 3, shown in Figure 8. In order to prevent the lift height of the robot's grinding device from exceeding its stroke distance, the top edge of the blade should be monitored to prevent the grinding device from sticking and causing failure. In this paper, two laser sensors are included at different heights on the lifting manipulator. Laser sensor 3 is set horizontally to detect the horizontal distance between the grinding robot and the wind turbine blade. Laser sensor 4 is attached at the top of the grinding robot to detect the distance between the robot and the highest point of the blade.

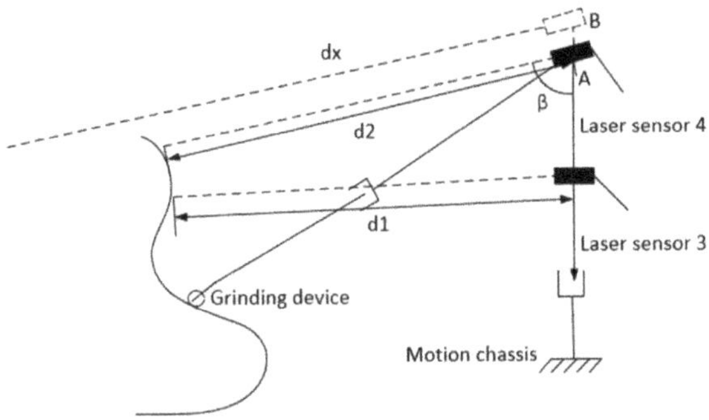

Figure 8. Detection of the blade's highest edge.

Here, d_1 is the vertical distance between the grinding robot and the real-time position of the blade, β is the angle of laser sensor 4, d_2 is the distance measured by laser sensor 2, and d_x is the distance as roughly calculated from angle β and d_1, the relevant formula for which is

$$d_x = \frac{d_1}{\sin\beta} \tag{3}$$

d_x as calculated by Equation (3) is compared with the distance d_2 measured by laser sensor 4. If d_2 is much larger than d_x, the grinding robot has reached the highest edge of the wind turbine blade; when d_2 gives the distance from laser sensor 2 to the ground, then the lift manipulator is commanded to stop moving and rise no further.

Figure 9 shows the control process of ensuring the grinding robot's safe positioning. Firstly, it is determined whether the vertical distance d_1 between the robot and the blade, as measured by laser sensor 3, is within the safe distance of 1.5~2.5 m. If it is within this range, the chassis stops moving. Then the distance d_2 is compared with d_x, and if $d_2 \gg d_x$, then the lift manipulator stops rising; otherwise, it commences the grinding action. If $d_1 \leq 1.5$ m, the robot is alerted and will stop moving forward. If $d_1 \geq 2.5$ m, the robot will continue to move closer to the wind turbine blade until it reaches a safe distance.

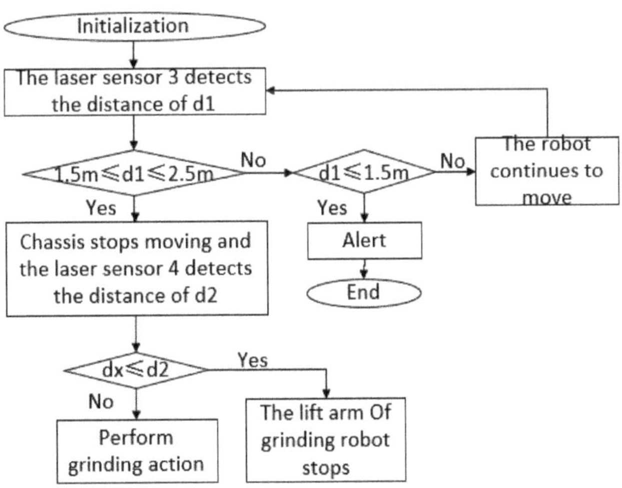

Figure 9. Flowchart showing the safe position control of the grinding robot.

3.2. Constant Force Grinding Control Strategy

According to the above analysis, the strategy used for controlling the grinding force and feed speed of the device is the key to realizing the constant control and compliance of the force of grinding. There are five degrees of freedom between the manipulator and the grinding device of the robot. The synchronous control strategy is illustrated in Figure 10, the ideal output of which is the hydraulic cylinder that controls the driving manipulator and the servo motor that drives the grinding device. In order to improve the response speed of the hydraulic system, fuzzy PID is used to control each hydraulic cylinder in the manipulator.

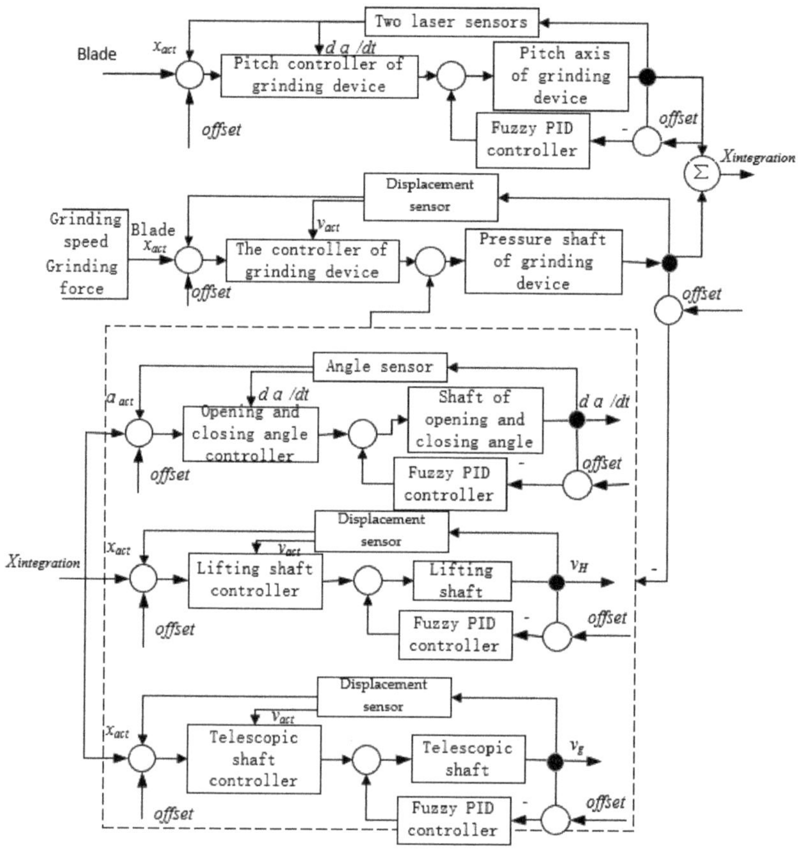

Figure 10. Five-axis synchronous control strategy diagram.

In the control strategy, the input variable used to control the grinding device is its tangential angle from the surface being processed, and the grinding speed and direction of the grinding force are the final outputs. The input variables used to control the grinding force are the blade parameters and the preset grinding force, and the final output is the grinding speed and the grinding force. The control of the input variable related to the manipulator is the final output of the former two inputs; the controller outputs the running speed of each part of the manipulator in order to maintain the grinding force and grinding speed required. Using the methods described above, the relevant control strategies for adjusting the grinding speed and the direction of grinding force are adjusted. Figure 11 shows the analysis method used to control the grinding speed and the grinding force.

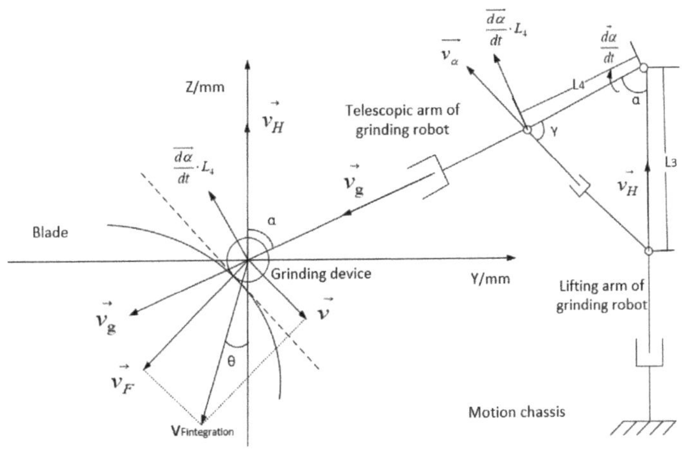

Figure 11. Diagram of the control strategy for grinding force and speed.

Here, v_H and v_g are the lifting speed of the lifting arm and the telescopic speed of the telescopic arm, respectively, and v_α is the speed of the hydraulic cylinder used for angle adjustment, which is not equal to the actual angular rotational velocity of the manipulator. Their corresponding relationship is

$$\frac{d\alpha}{dt}L_4 = v_\alpha \sin\gamma \tag{4}$$

where $\frac{d\alpha}{dt}$ is the angular velocity of the opening and closing of the mechanical arm and represents the actual speed of the grinding device; L_3 and L_4, respectively, represent the distances between the installed position of the hydraulic cylinder used for angle adjustment and the rotational center of the manipulator being adjusted; α is the angle between the lifting and telescopic manipulators; and γ is the angle between the telescopic manipulator and the hydraulic cylinder used to adjust the angle. α can be measured in real time by the angle sensor, and γ can be obtained via the following relationship:

$$\gamma = \arctan\left(\frac{L_3 \sin\alpha}{L_4 - L_3 \cos\alpha}\right) \tag{5}$$

Combining Equations (4) and (5), to meet the required opening and closing speed of the manipulator, the speed provided by the hydraulic cylinder for angle adjustment must be

$$v_\alpha = \frac{d\alpha}{dt} \cdot \frac{L_4}{\sin\left[\arctan\left(\frac{L_3 \sin\alpha}{L_4 - L_3 \cos\alpha}\right)\right]} \tag{6}$$

The grinding device adjusts the pressure applied on the blade by changing the displacement l, which is measured by the micro displacement sensor installed in the pose adjustment device. The resultant vector of the differentiated feed speed v_F and the stretching speed of the grinding device must both be adjusted in real time to ensure the correct grinding force, as shown by the following equation:

$$\overrightarrow{v_{Fintegration}} = \overrightarrow{v_F} + \vec{v} = \frac{\overrightarrow{dl}}{dt} + \vec{v} \tag{7}$$

The mathematical relationship between the coordinate systems of the grinding robot and those of each joint is essentially nonlinear, and can be expressed by the following nonlinear vector value function:

$$\mathrm{x}(t) = f[q(t)] \tag{8}$$

So, the operating speed ultimately required is given by

$$\dot{x} = \left[\dot{x_1}, \dot{x_2}, \cdots \dot{x_m}\right]^T = v_{integration} \tag{9}$$

To provide the speed ultimately required, the speed required from each joint is

$$\dot{q} = \left[\dot{q}_1, \dot{q}_2, \cdots \dot{q}_n\right]^T \tag{10}$$

The relationship between the required speed of the end and the speed of each link can be obtained by applying a kinematic calculation to the robot, which is shown in the following expression:

$$\dot{x} = J\dot{q} \tag{11}$$

where J is the Jacobi matrix of the robot, the specific value of which can be derived from the kinematics of the grinding robot, as shown in the following expression:

$${}^{T}J(q) = \left[{}^{T}J_1(q)\ {}^{T}J_2(q)\ {}^{T}J_3(q)\ {}^{T}J_4(q)\ {}^{T}J_5(q)\right] \tag{12}$$

The elements in the formula are

$$\begin{gathered}{}^{T}J_1(q) = \begin{bmatrix} 0 \\ (d_2 - d_4c\theta_3)(c\theta_5s\theta_3 - c\theta_3s\theta_5) + (d_4s\theta_3 + a_2)(s\theta_3s\theta_5 - c\theta_3c\theta_5) \\ d_4s\theta_3 + a_2 \\ c\theta_5s\theta_3 - c\theta_3s\theta_5 \\ -s\theta_3s\theta_5 - c\theta_3c\theta_5 \\ 0 \end{bmatrix} \\ {}^{T}J_2(q) = \begin{bmatrix} 0 \\ 0 \\ -1 \\ 0 \\ 0 \\ 0 \end{bmatrix} {}^{T}J_3(q) = \begin{bmatrix} 0 \\ -d_4s\theta_5 \\ 0 \\ s\theta_5 \\ c\theta_5 \\ 0 \end{bmatrix} {}^{T}J_4(q) = \begin{bmatrix} 0 \\ 0 \\ 1 \\ 0 \\ 0 \\ 0 \end{bmatrix} {}^{T}J_5(q) = \begin{bmatrix} 0 \\ 0 \\ 0 \\ 0 \\ 0 \\ 1 \end{bmatrix}\end{gathered} \tag{13}$$

The above-designed control strategy ensures that the correct speed is provided by each connecting rod at any given time in order to meet the grinding requirements, as

$$q(t) = J^{-1}(q)\dot{q}(t) \tag{14}$$

In order to resolve the problem whereby the inverse Jacobian matrix of the grinding robot does not exist because of the redundancy of the mechanical arm, the following expression can be used [31]:

$$C = \frac{1}{2}\dot{q}^T A\dot{q} + \lambda^T\left[\dot{x} - J(q)\dot{q}\right] \tag{15}$$

where C is an m × m symmetric positive definite matrix, and λ is the cost judgement. By minimizing C, the speed of each link and λ, we can obtain the following expression:

$$\dot{q}(t) = A^{-1}J^T(q)\lambda \tag{16}$$

$$\dot{q}(t) = J(q)\dot{q}(t) \tag{17}$$

By combining the above two expressions, λ can be derived as

$$\lambda = [J(q)A^{-1}J^T(q)]^{-1}\dot{x}(t) \tag{18}$$

By combining Equations (16) and (18), the desired speed $x(t)$ is given for the end of the grinding device, and the speed of each joint is obtained according to Equation (18):

$$\dot{q}(t) = A^{-1}J^T(q)[J(q)A^{-1}J^T(q)]^{-1}\dot{x}(t) \tag{19}$$

where A is an identity matrix. Equation (19) can be simplified to Equation (14), which relates the grinding blade's position and time. The position of the end of the grinding device, which is related to the sample taken at any given time, is a fixed value, and the sampling period is 200 ms.

Due to the redundancy of the grinding robot, there are multiple solutions to determining the expected speed of the robot end at any time. In order to ensure optimal control of the robot, the change in speed of the connecting rod must be as small as possible in any sampling period. Using the sensor and a calculation of the mathematical relationship, the angle between the desired direction of travel at the end and the direction of each connecting rod at any given time can be obtained, as shown in Figure 12.

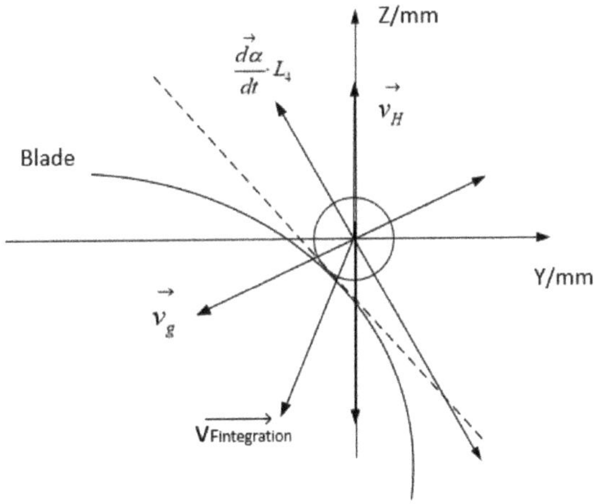

Figure 12. Instantaneous velocity relationship of contact force control strategy.

Give its three-part structure, the robot can be moving in two directions at any given time. For example, the lifting system can be raising or lowering at any time. In order to ensure its compliance to control, the angle between its direction of motion and the desired direction of the end must be as small as possible [32]. The speeds required of each connecting rod are calculated by Equation (19). On this basis, the required driving force, or torque, which is also the force provided by the hydraulic cylinder at any time, can be obtained by the following expression:

$$f_4 = D_{44}\ddot{q}_4(t) + D_{45}\ddot{q}_5(t) + D_{433}\dot{q}_3(t)\dot{q}_4(t) \tag{20}$$

where $D_{ijk}(q)$ is the inertial matrix of the grinding robot.

3.3. Control Strategy of Mechanical Arm Power System

In maintaining constant control over the force of the grinding robot, the position of the manipulator will change at any given time, so the PID parameters of the hydraulic control system need to be dynamically adjusted in real time. Based on the necessity of real-time surface tracking of the turbine blade and the application of constant force during grinding, the fuzzy PID control principle used to control the hydraulic system is shown in Figure 13.

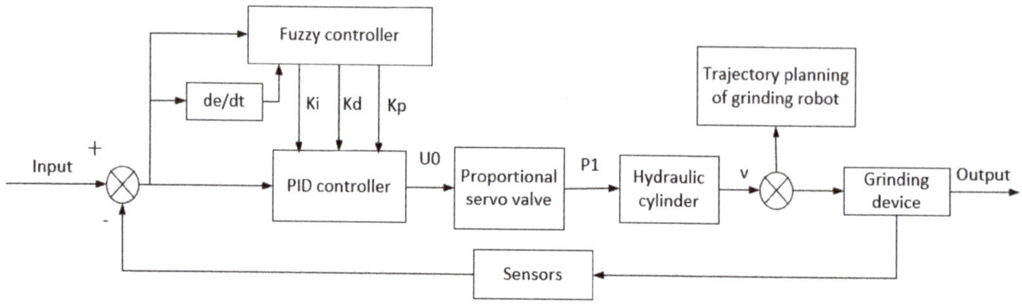

Figure 13. Fuzzy PID control principle.

Through the analysis of the grinding control system, it is finally determined that the inputs of the fuzzy PID control are the deviation and deviation increment of the combined vector of the expected grinding feed speed and the expected grinding force and the combined vector of the two vectors at any time, which is used as the input to control the adjustment of the fuzzy parameters K_p, K_i and K_d in real time.

3.3.1. Fuzzy Input Value

In order to achieve fuzzification, it is necessary to fuzzify the calculated values. In this design, the fuzzy subset is {NB, NM, NS, ZO, PS, PM, PB}, the domain is $\{-6, -5, -4, -3, -2, -1, 0, 1, 2, 3, 4, 5, 6\}$. There is an interval for the measured force and velocity signals, which is made as same as the PID adjustment time. Deviation and deviation change rate are shown as:

$$f(e) = \frac{6 \cdot e}{V_{max} - V_{min}} \tag{21}$$

$$f(e) = \frac{6 \cdot ec}{2(V_{max} - V_{min})} \tag{22}$$

where e is the deviation and ec is the deviation increment. The quantization of input error and error change rate can be obtained by using the above equation. The functional relationship of the linear membership function is shown in Figure 14.

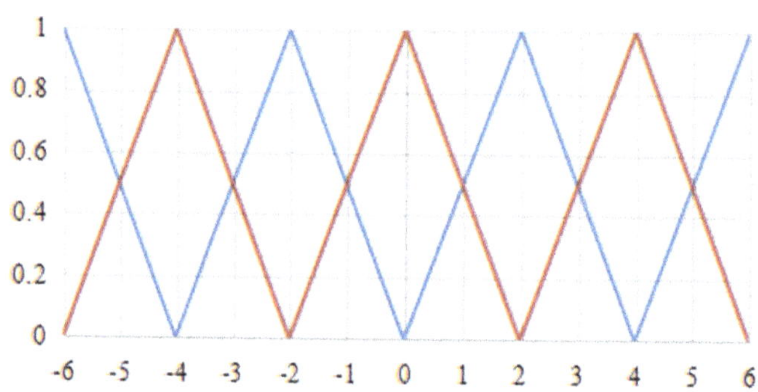

Figure 14. Membership function of fuzzy controller.

3.3.2. Establish Fuzzy Rule Table

After modeling the input, K_p, K_i and K_d in the PID control are adjusted. The specific fuzzy rules are shown in Table 1. According to the fuzzy rule table and the deviation and its change rate calculated by the value collected by the sensor and the expected value, the fuzzy subsets corresponding to ΔK_p, ΔK_i and ΔK_d are obtained.

Table 1. Fuzzy control rules table.

K_p K_i K_d		*ec*						
		NB	**NM**	**NS**	**ZO**	**PS**	**PM**	**PB**
e	NB	PB NB PS	PB NB NS	PM NM NB	PM NM NB	PS NS NB	ZO ZO NM	ZO ZO PS
	NM	PB NB PS	PB NB NS	PM NM NB	PS NS NM	PS NS NM	ZO ZO NS	NS ZO PS
	NS	PM NB ZO	PM NM NS	PM NS NM	PS NS NM	ZO ZO NS	NS PS NS	NS PS ZO
	ZO	PM NM ZO	PM NM NS	PS NS NS	ZO ZONS	NS PS NS	NM PM NS	NS PS ZO
	PS	PS NM ZO	PS NS ZO	ZO ZO ZO	NS PS ZO	NS PS ZO	NM PM ZO	NB PB ZO
	PM	PS ZO PB	ZO ZO NS	NS PS PS	NM PS PS	NM PM PS	NM PB PS	NB PB PB
	PB	ZO ZO PB	ZO ZO PM	NM PS PM	NM PM PM	NM PM PS	NB PB PS	NB PB PB

3.3.3. Demoulding Processing

Since K_p, K_i and K_d are adjusted in the actual fuzzy control, it can be quantified by using the center of gravity calculation according to the triangular membership function, as shown in Equation (23).

$$v_0 = \frac{\sum_{i=0}^{n} M_i F_i}{\sum_{i=0}^{n} M_i} \tag{23}$$

where M is membership and F is fuzzy quantization value. The sum of membership is 1, so the denominator can be omitted, so that the calculation of each object is actually a matrix operation:

$$K = [M_{e1}\ M_{e2}] \begin{bmatrix} F_a & F_b \\ F_e & F_d \end{bmatrix} [M_{ec1}\ M_{ec2}]^T \tag{24}$$

Coefficients can also be introduced to enlarge and reduce the variation of K_p, K_i and K_d after the increment is obtained.

$$K(n) = K(n-1) + \Delta K \cdot \lambda \tag{25}$$

where λ is set to 1 in this control strategy and ΔK is the calculated value. The hydraulic cylinder driving the lifting manipulator is taken as an example for this experimental analysis. The performance indices of the hydraulic cylinder with fuzzy PID control and without fuzzy PID control are compared, as shown in Figure 15, and their specific indices are shown in Table 2.

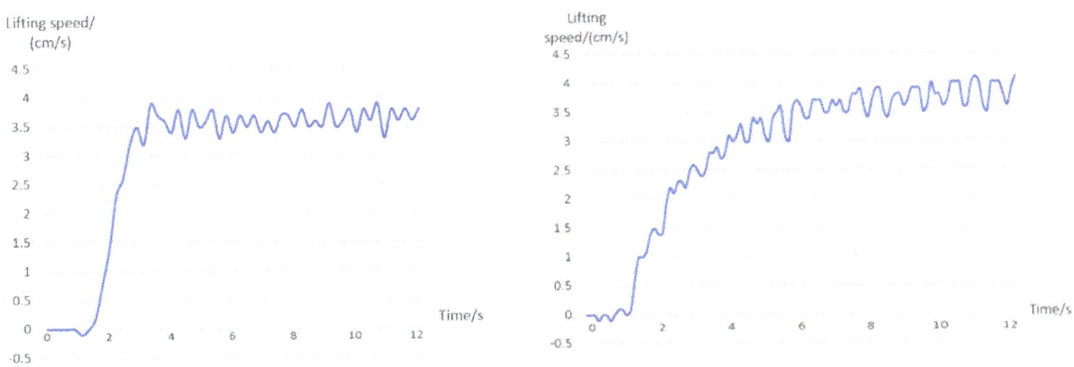

(**a**) (**b**)

Figure 15. The performance indices. (**a**) with fuzzy PID control and (**b**) without fuzzy PID control.

Table 2. Comparison of specific indicators of hydraulic cylinders with different control methods.

Control Mode	Maximum Overshoot	Rise Time	Peak Time	Steady State Error	Accommodation Time
fuzzy PID control	3.58%	700 ms	1500 ms	± 0.5 cm/s	3100 ms
proportional servo valve control	5.58%	1500 ms	2700 ms	± 2 cm/s	5300 ms

It can be seen that, when using fuzzy PID control, the hydraulic cylinder shows a small degree of overshoot and fast response, and the errors arising from the designed static control strategy can be greatly reduced.

3.4. Analysis of Gravity Compensation

The grinding device of the robot described in this paper is attached to the end of the manipulator by a fixed connection, and a schematic diagram of the robot's coordinates of movement is shown in Figure 16. In addition to the parameters relevant to the control strategy, the factors influencing the stability of the control strategy also include the contribution of the grinding device's own gravity to the force applied.

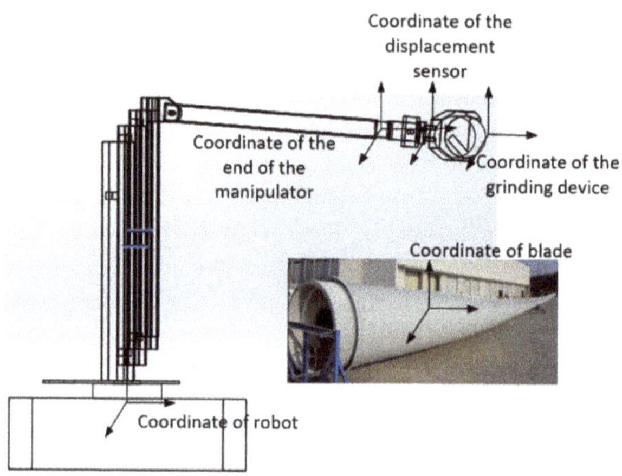

Figure 16. Diagram of relative positional coordinates.

The overall force includes the contact force F_E, the gravity of the grinding device G_T and the inertial force of acceleration during the movement of the manipulator F_I, so the force measured by the sensor cannot be directly used in the algorithm. The influence of inertial force can be ignored because the acceleration of the manipulator is negligible, and gravity compensation can be incorporated to eliminate the influence of the grinding device's gravity. Figure 17 shows the analysis of the force measured by the sensor in the grinding device, and the relation is shown in Equation (26).

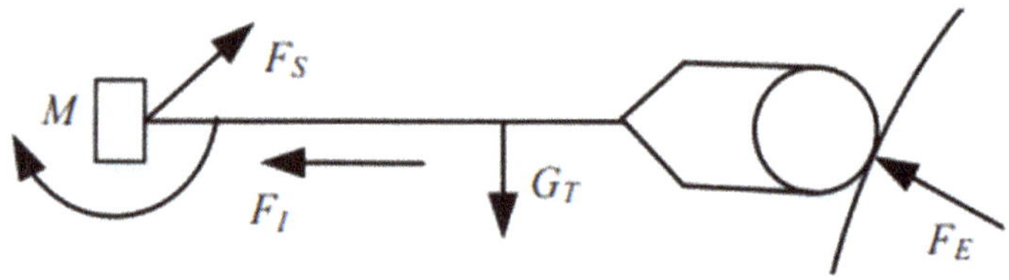

Figure 17. Analysis diagram of force measured by sensor.

$$\overrightarrow{F_S} = \overrightarrow{F_E} + \overrightarrow{G_T} + \overrightarrow{F_I} \quad (26)$$

This can be simplified as

$$\overrightarrow{F_S} = \overrightarrow{F_E} + \overrightarrow{G_T} \tag{27}$$

Using the rotation transformation matrix of the robot, the force of the gravity of the grinding device as measured by the sensor can be expressed as

$$^{S}G_T = {}^{S}_{E}R\,{}^{E}_{O}R\,{}^{O}G_T \tag{28}$$

where $^{S}G_T$ is the gravitational force vector of the front-end grinding device in the coordinate system given by the displacement sensor, ${}^{S}_{E}R$ is the rotation matrix derived from the system of axes of the terminal part of the robotic arm in relation to the sensor's coordinate system and ${}^{E}_{O}R$ is the rotation matrix of the frame of the axes of the grinding device in relation to the basal coordinate system, the relative expression of which is

$${}^{E}_{O}R = \begin{bmatrix} c\theta_1(c\theta_3 c\theta_5 + s\theta_3 s\theta_5) & c\theta_1(s\theta_3 s\theta_5 - c\theta_3 s\theta_5) & -s\theta_1 \\ s\theta_1(s\theta_3 c\theta_5 + s\theta_3 s\theta_5) & s\theta_1(s\theta_3 c\theta_5 - c\theta_3 s\theta_5) & c\theta_1 \\ s\theta_3 c\theta_5 - c\theta_3 s\theta_5 & -s\theta_3 s\theta_5 - c\theta_3 s\theta_5 & 0 \end{bmatrix} \tag{29}$$

where θ_i (i = 1~5) is the angle of the joints between mechanical arms I to V. There is a fixed deflection θ between the fixed system of axes of the grinding device and the system of axes of the sensor used to measure the pressure, as shown in Figure 18.

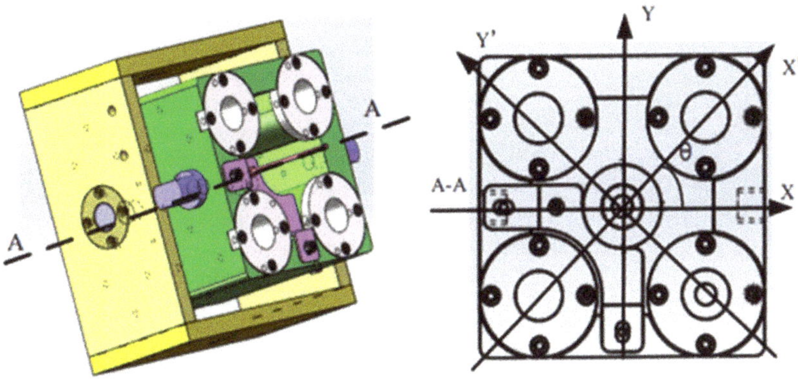

Figure 18. The grinding device and the sensor's coordinate systems.

The rotational transformation of the system of axes of the sensor relative to the flange's coordinate system is

$${}^{S}_{E}R = \begin{bmatrix} c\theta & -s\theta & 0 \\ s\theta & c\theta & 0 \\ 0 & 0 & 1 \end{bmatrix} \tag{30}$$

Combining the above four formulas, the following expression can be obtained:

$$\begin{bmatrix} F_x \\ F_y \\ F_z \end{bmatrix} = \begin{bmatrix} c\theta & -s\theta & 0 \\ s\theta & c\theta & 0 \\ 0 & 0 & 1 \end{bmatrix} \cdot \begin{bmatrix} c\theta_1(c\theta_3 c\theta_5 + s\theta_3 s\theta_5) & c\theta_1(s\theta_3 s\theta_5 - c\theta_3 s\theta_5) & -s\theta_1 \\ s\theta_1(s\theta_3 c\theta_5 + s\theta_3 s\theta_5) & s\theta_1(s\theta_3 c\theta_5 - c\theta_3 s\theta_5) & c\theta_1 \\ s\theta_3 c\theta_5 - c\theta_3 s\theta_5 & -s\theta_3 s\theta_5 - c\theta_3 s\theta_5 & 0 \end{bmatrix} \cdot \begin{bmatrix} 0 \\ 0 \\ -G_T \end{bmatrix} \tag{31}$$

Simplifying Equation (31), the following expression can be given:

$$cos\theta = \sqrt{\frac{F_x^2 + cos^2(\theta_3 + \theta_5)}{cos^2(\theta_3 + \theta_5) + sin^2(\theta_3 - \theta_5)}} \tag{32}$$

Multiple sets of data pertaining to the installation deflection angle θ can be derived from information on the angles of the robot's different poses, and an average value can be obtained. Because the micro displacement sensor is used to collect force information, the force that is measured is only unidirectional, and its calculation formula is

$$^{s}F_E = F - \frac{G_T cos(\theta_3 - \theta_5) cos(\theta + \theta_1) / cos^2\theta_5}{tan^2\theta_5 + (cos^2\theta_3 + sin\ \theta_3 sin\ \theta_1) + cos\ \theta_3 tan\ \theta_5 (sin\ \theta_3 - sin\ \theta_1)} \quad (33)$$

where $^{s}F_E$ is the grinding force as a component of the coordinate system measured by the pressure sensor, and F is the grinding force calculated by the displacement sensor. In the process of grinding a given area, the rotational angle of the grinding robot is always $\pm 90°$, so Equation (34) can be simplified as

$$^{s}F_E = F - \frac{G_T cos(\theta_3 - \theta_5) sin\theta / cos^2\theta_5}{tan^2\theta_5 + cos^2\theta_3 + cos\ \theta_3 tan\ \theta_5 sin\ \theta_3} \quad (34)$$

The operational flow of gravity compensation in the grinding robot is as follows:

- The grinding device performs a profiling movement on the blade's surface when it is not in contact with the blade and the grinding force is zero, after which time the measurements of the pressure sensor and the manipulator in each attitude are collected;
- The force and attitude information collected are substituted into Equation (33) to obtain the installation deflection angle;
- The above two steps are repeated to collect multiple different sets of information and thus obtain multiple installation deflection angles, which can be used to calculate the average installation deflection angle θ;
- The deflection angle obtained in the previous step is substituted into Equation (34) to determine whether the result is 0; if it is not 0, the above steps are repeated until the contact force obtained is 0.

In this way, gravity compensation can be realized; the influence of the gravitational force of the grinding device can be eliminated using the above formula, and the real contact force can thus be obtained, which can then be applied in the constant force control algorithm.

4. Experiments

To verify the feasibility of our control strategy, the surface tracking and constant force control of the grinding robot have been experimentally verified in a grinding operation. The experimental results of robotic grinding are here compared with the artificial results, and this verifies the feasibility of the designed control strategy, according to which the grinding robot can meet the requirements of a given processing site.

4.1. Control Algorithm Comparison Experiment

Taking the driving hydraulic cylinder at the opening and closing manipulator as an example, the experimental analysis is carried out. The performance indices of hydraulic cylinders using fuzzy PID control and normal PID control when the length of the telescopic arm is 218.5 mm are compared as shown in Figure 19. The maximum overshoot is 0.27 rad/s.

The reason for the large difference between the two response curves is that the opening and closing angle corresponds to the expansion length of the hydraulic cylinder. Normal PID has different control effects when the hydraulic cylinder telescopic length is different. However, the three parameters K_p, K_i and K_d of fuzzy PID are variable, which can adapt to the different lengths of hydraulic cylinder. The experiment shows that fuzzy PID control is more adaptable and superior than normal PID control.

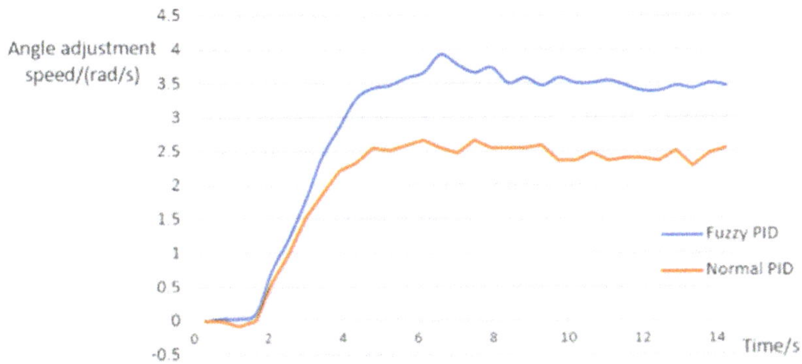

Figure 19. Curve of velocity response.

4.2. *Surface Tracking Experiment*

This surface tracking experiment uses time intervals of 200 ms. The coordinate sizes of the tangent point of the grinding device and the blade relative to the base coordinates of the grinding robot are calculated using data given by the five laser sensors installed on the manipulator, enabling us to monitor the actual trajectory of the blade.

In a given grinding area, the grinding robot does not change direction on the x-axis, and this value is always 0. The formulae for calculating the coordinates of the tangent point of the robot and the wind turbine blade in the y-direction and z-direction are

$$\mathrm{y} = d_4 \sin\left(\frac{\pi\theta_3}{180^{\circ}}\right) + 600\cos\left(\pi\frac{\theta_3+\theta_5-180^{\circ}}{180^{\circ}}\right) \quad (35)$$

$$\mathrm{z} = d_2 - d_4 \cos\left(\frac{\pi\theta_3}{180^{\circ}}\right) + 600\sin\left(\pi\frac{\theta_3+\theta_5-180^{\circ}}{180^{\circ}}\right) \quad (36)$$

where d_2 is the position of the lifting manipulator, d_4 is the telescopic length of the telescopic manipulator, θ_3 is the opening and closing angle of the mechanical arms, and θ_5 is the pitch angle of the polishing device. The calculated trajectory of the cutting point coordinates in a grinding area is shown in Figure 20, and they are shown to be very close to the contours of the processing area, which verifies the utility of this control strategy for surface tracking during grinding.

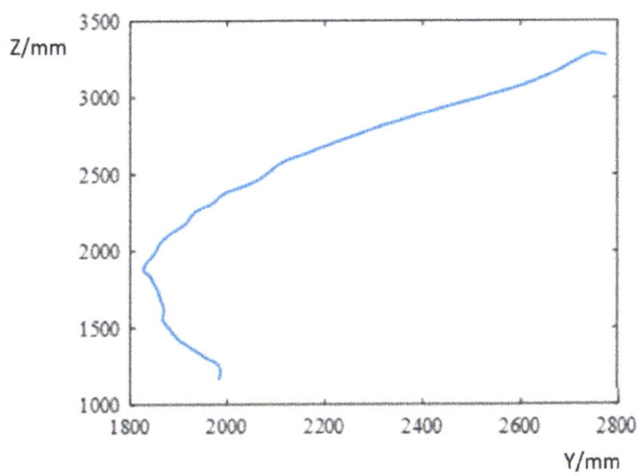

Figure 20. Trajectory of grinding cutting point.

4.3. Grinding Force Tracking Experiment

The expected values of the grinding force and feed speed are set to 1500 N and 150 mm/s, respectively, and these are compared with data collected for a grinding area at sampling periods of 200 ms. Other factors affecting the grinding are set to constant values, such as the revolution speed of the grinding device, which is 1500 r/min. Figure 21 shows the grinding contact force and feed speed, collected from top to bottom during the grinding of a processing area. The grinding contact force is maintained at about 1500 N, and the grinding feed speed remains basically stable at about 150 mm/s after contact, which verifies the utility of our strategy.

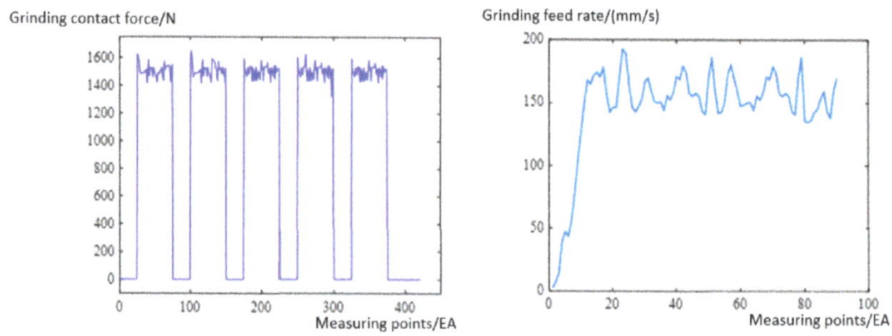

Figure 21. Feedback values of grinding force and grinding feed speed.

4.4. Field Processing Experiment

Actual processing experiments of the grinding robot also needed to be carried out to verify that it could meet the needs of field processing. In the artificial grinding area and the robot grinding area, two square areas with side lengths of 500 mm were randomly selected. Within these areas, detection points were set every 100 mm along the axial direction of the grinding device and the grinding feed direction to measure the roughness of the machined surface, and the measured data are shown in Figure 22. The average roughness yielded by artificial grinding was 5.0934 μm. The average roughness yielded by robot grinding was 2.4828 μm, and all measurement points yielded data that met the surface roughness requirements for an optimal subsequent painting process.

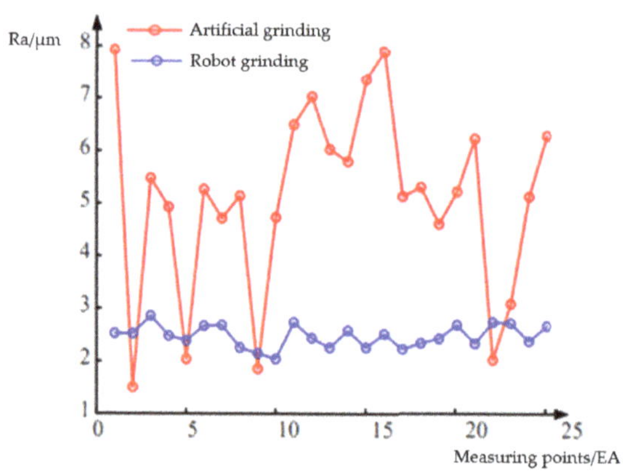

Figure 22. Grinding surface roughness.

5. Discussion

In this paper, we introduced a robot that can be used for grinding the surface of wind turbine blades, and we proposed related control and motion planning algorithms; we found that tracking the blade's surface and employing the constant force control algorithm improved the grinding effect. Fuzzy PID control of the power source, focusing on the hydraulic cylinder of the manipulator, was carried out, and this greatly improved the response speed and reduced errors in the static control strategy. In order to eliminate the impact of the weight of the grinding device on the information collected regarding grinding force, a gravity compensation method for the grinding device was designed.

The experiments show that fuzzy PID has excellent adaptability and can meet the design requirements. The effect of tracking the blade's surface during the grinding process is very impressive, enabling the average grinding force and grinding feed speed to be maintained at near the predicted values. In the actual processing experiment, the surface qualities of blades after robot grinding and artificial grinding were compared. The mean roughness of the blade after grinding by robot was Ra = 2.4828 μm, which is within the optimal range of Ra = 2.0~3.0 μm, and this makes the quality of the blade surface benefit the subsequent painting process.

Future work should study the joint motion of the chassis and the manipulator of the grinding robot. In addition to the grinding force and feed speed, other factors that influence the grinding effect, such as the grinding speed and the amount of sandpaper in the grinding device, can also be further studied.

Author Contributions: Conceptualization, X.S.; methodology, X.S. and S.F.; software, S.F.; validation, M.L., Y.D. and S.F.; formal analysis, Y.D.; investigation, M.L.; resources, X.S.; data curation, Y.D.; writing—original draft preparation, M.L. and S.F.; writing—review and editing, X.S., M.L. and Y.D.; visualization, S.F.; supervision, X.S.; project administration, X.S.; funding acquisition, X.S. All authors have read and agreed to the published version of the manuscript.

Funding: This research received no external funding.

Institutional Review Board Statement: Not applicable.

Informed Consent Statement: Not applicable.

Data Availability Statement: No new data were created.

Conflicts of Interest: The authors declare no conflict of interest.

References

1. Dai, J.C.; Yang, X.; Wen, L. Development of wind power industry in China: A comprehensive assessment. *Renew. Sustain. Energy Rev.* **2018**, *97*, 156–164. [CrossRef]
2. Liu, Y.D.; Wang, J. The SWOT Analysis and Countermeasure Research on the Development of Wind Power Industry in China. In Proceedings of the IOP Conference Series: Earth and Environmental Science, Chengdu, China, 9–11 July 2021.
3. Chen, G.W.; Huang, X.M.; Zhang, L.A.; Wang, J.H.; Liu, W.S. Study on sand grinding technology of wind turbine blade surface. *Acta Energ. Sol. Sin.* **2021**, *42*, 302–307.
4. Yan, C.; Chen, X.L.; Li, G.L.; Li, C.L. Design and development of an efficient and adaptive grinding head system for a new type of wind turbine blade automatic grinding robot. *Compos. Sci. Eng.* **2021**, *12*, 114–119.
5. Xu, Z.L.; Lu, S.; Yang, J.; Feng, Y.H.; Shen, C.T. A wheel-type in-pipe robot for grinding weld beads. *Adv. Manuf.* **2017**, *5*, 182–190. [CrossRef]
6. Ma, W.C. Design and Research on Constant Force Control Device for Robot Grinding. Master's Thesis, Huazhong University of Science & Technology, Wuhan, China, May 2019.
7. Xiao, M. Research on Constant Force Control Methods in Robot Grinding Process. Ph.D. Thesis, South China University of Technology, Guangzhou, China, May 2020.
8. Zhang, T.; Yu, Y.; Zou, Y.B. An Adaptive Sliding-Mode Iterative Constant-force Control Method for Robotic Belt Grinding Based on a One-Dimensional Force Sensor. *Sensors* **2019**, *19*, 1635. [CrossRef] [PubMed]
9. Kuo, Y.L.; Huang, S.Y.; Lan, C.C. Sensorless force control of automated grinding/deburring using an adjustable force regulation mechanism. *IEEE Int. Conf. Robot. Autom.* **2019**, *2019*, 9489–9495.
10. Shen, Y.; Lu, Y.; Zhuang, C. A fuzzy-based impedance control for force tracking in unknown environment. *J. Mech. Sci. Technol.* **2022**, *36*, 5231–5242. [CrossRef]
11. Wahballa, H.; Duan, J.; Dai, Z. Constant force tracking using online stiffness and reverse damping force of variable impedance controller for robotic polishing. *Int. J. Adv. Manuf. Technol.* **2022**, *121*, 5855–5872. [CrossRef]
12. Yao, M.Y.; Zheng, C.W.; Wang, Z.H.; Qian, J.Z.; Xi, Q.; Kuang, S.L. Research on the method of skateboard edge grinding by combination of robot and pneumatic constant force grinding device. In Proceedings of the Fourth International Conference on Mechanical, Electric and Industrial Engineering (MEIE2021), Kunming, China, 22–24 May 2021.

13. Sun, M.J.; Guo, K.; Sun, J. Research on robot grinding force control method. *Intell. Robot. Appl.* **2021**, *13014*, 821–829.
14. Dai, S.J.; Li, S.N.; Ji, W.B.; Sun, Z.L.; Zhao, Y.F. Force tracking control of grinding end effector based on backstepping + PID. *Ind. Robot. Int. J.* **2021**, *49*, 34–46. [CrossRef]
15. Xu, X.H.; Chen, W. Hybrid active/passive force control strategy for grinding marks suppression and profile accuracy enhancement in robotic belt grinding of turbine blade. *Robot. Comput. Integr. Manuf.* **2021**, *67*, 102047. [CrossRef]
16. Zhang, T.; Yuan, C.; Zou, Y.B. Online Optimization Method of Controller Parameters for Robot Constant Force Grinding Based on Deep Reinforcement Learning Rainbow. *J. Intell. Robot. Syst.* **2022**, *105*, 85. [CrossRef]
17. Zhao, W.; Xiao, J.L.; Liu, S.J.; Dou, S.X.; Liu, H.T. Robotic direct grinding for unknown workpiece contour based on adaptive constant force control and human–robot collaboration. *Ind. Robot.* **2022**, *50*, 376–384. [CrossRef]
18. Wang, G.; Deng, Y.; Zhou, H.; Yue, X. PD-adaptive variable impedance constant force control of macro-mini robot for compliant grinding and polishing. *Int. J. Adv. Manuf. Technol.* **2023**, *124*, 2149–2170. [CrossRef]
19. Han, J.L.; Shan, X.L.; Liu, H.T.; Xiao, J.L.; Huang, T. Fuzzy gain scheduling PID control of a hybrid robot based on dynamic characteristics. *Mech. Mach. Theory* **2023**, *184*, 105283. [CrossRef]
20. Li, D.W.; Yang, J.X.; Zhao, H.; Ding, H. Contact force plan and control of robotic grinding towards ensuring contour accuracy of curved surfaces. *Int. J. Mech. Sci.* **2022**, *227*, 107449. [CrossRef]
21. Jia, W.D.; Jiang, Z.F.; Dai, Y. Constant Force Control Method of Grinding Device. *Lect. Notes Comput. Sci.* **2022**, *13457*, 378–387.
22. Zhao, X.W.; Lu, H.; Yu, W.F.; Tao, B.; Ding, H. Vision-based Mobile Robotic Grinding for Large-scale Workpiece and Its Accuracy Analysis. *IEEE/ASME Trans. Mechatron.* **2022**, *28*, 895–906. [CrossRef]
23. Dai, S.J.; Zhang, W.H.; Ji, W.B.; Zhao, Y.F.; Zheng, H.W.; Mu, J.H.; Li, P.W.; Deng, R.Q. Research on constant force grinding control of aero-engine blades based on extended state observer. *Ind. Robot* **2022**, *49*, 1077–1088. [CrossRef]
24. Xiao, M.; Zhang, T.; Zou, Y.B.; Chen, S.Y. Robotic constant force grinding control based on grinding model and iterative algorithm. *Ind. Robot* **2020**, *48*, 270–279. [CrossRef]
25. Zhang, X.X. Hybrid Coutour Force/Position Control of Industrial Robot Based on Variable-Gain Iterative Learning. Ph.D. Thesis, China University of Geosciences, Wuhan, China, May 2022.
26. Li, J.; Guan, Y.; Chen, H.; Wang, B.; Zhang, T. Robotic Polishing of Unknown-Model Workpieces With Constant Normal Contact Force Control. *IEEE/ASME Trans. Mechatron.* **2023**, *28*, 1093–1103. [CrossRef]
27. Zhang, T.; Yuan, C.; Zou, Y.B. Research on the algorithm of constant force grinding controller based on reinforcement learning PPO. *Int. J. Adv. Manuf. Technol.* **2023**, 1–14. [CrossRef]
28. Peterson, M.B. Design considerations for effective wear control. In *Wear Control Handbook*; American Society of Mechanical Engineering: New York, NY, USA, 1980; pp. 413–474.
29. Tsai, M.J.; Huang, J.F.; Kao, W.L. Robotic polishing of precision molds with uniform material removal control. *Int. J. Mach. Tools Manuf.* **2009**, *49*, 885–895. [CrossRef]
30. Lai, H.S.; Wu, L.; Chen, X.D.; Yang, Z.Q. Research on the force control of the robot and surface tracking with unknown parameters. *Manuf. Technol. Mach. Tool* **2018**, *6*, 21–25.
31. Xiong, M.Q. Research on Key Techniques of Robot Force/Position Control Grinding and Polishing. Master's Thesis, Changchun University of Science and Technology, Changchun, China, June 2020.
32. Zhang, T.; Xiao, M.; Zou, Y.B.; Xiao, J.D. Research on robot constant force control of surface tracking based on reinforcement learning. *J. Zhejiang Univ. Eng. Sci.* **2019**, *53*, 1865–1873. [CrossRef]

sensors

Article

Robotic Intracellular Pressure Measurement Using Micropipette Electrode

Minghui Li [1,2], Jinyu Qiu [1,2], Ruimin Li [1,2], Yuzhu Liu [1,2], Yue Du [1,2], Yaowei Liu [1,2], Mingzhu Sun [1,2], Xin Zhao [1,2] and Qili Zhao [1,2,3,*]

1 Institute of Robotics and Automatic Information System, Tianjin Key Laboratory of Intelligent Robotics, Nankai University, Tianjin 300350, China; 2120210388@mail.nankai.edu.cn (M.L.); qiujinyu@mail.nankai.edu.cn (J.Q.); lrumin@mail.nankai.edu.cn (R.L.); liuyuzhu@mail.nankai.edu.cn (Y.L.); duy@nankai.edu.cn (Y.D.); liuyaowei@mail.nankai.edu.cn (Y.L.); sunmz@nankai.edu.cn (M.S.); zhaoxin@nankai.edu.cn (X.Z.)

2 Institute of Intelligence Technology and Robotic Systems, Shenzhen Research Institute of Nankai University, Shenzhen 518083, China

3 Beijing Advanced Innovation Center for Intelligent Robots and Systems, Beijing Institute of Technology, Beijing 100081, China

* Correspondence: zhaoqili@nankai.edu.cn

Abstract: Intracellular pressure, a key physical parameter of the intracellular environment, has been found to regulate multiple cell physiological activities and impact cell micromanipulation results. The intracellular pressure may reveal the mechanism of these cells' physiological activities or improve the micro-manipulation accuracy for cells. The involvement of specialized and expensive devices and the significant damage to cell viability that the current intracellular pressure measurement methods cause significantly limit their wide applications. This paper proposes a robotic intracellular pressure measurement method using a traditional micropipette electrode system setup. First, the measured resistance of the micropipette inside the culture medium is modeled to analyze its variation trend when the pressure inside the micropipette increases. Then, the concentration of KCl solution filled inside the micropipette electrode that is suitable for intracellular pressure measurement is determined according to the tested electrode resistance–pressure relationship; 1 mol/L KCl solution is our final choice. Further, the measurement resistance of the micropipette electrode inside the cell is modeled to measure the intracellular pressure through the difference in key pressure before and after the release of the intracellular pressure. Based on the above work, a robotic measurement procedure of the intracellular pressure is established based on a traditional micropipette electrode system. The experimental results on porcine oocytes demonstrate that the proposed method can operate on cells at an average speed of 20~40 cells/day with measurement efficiency comparable to the related work. The average repeated error of the relationship between the measured electrode resistance and the pressure inside the micropipette electrode is less than 5%, and no observable intracellular pressure leakage was found during the measurement process, both guaranteeing the measurement accuracy of intracellular pressure. The measured results of the porcine oocytes are in accordance with those reported in related work. Moreover, a 90% survival rate of operated oocytes was obtained after measurement, proving limited damage to cell viability. Our method does not rely on expensive instruments and is conducive to promotion in daily laboratories.

Keywords: intracellular pressure; electrode circuit model; micropipette electrode; micro-nano operation

Citation: Li, M.; Qiu, J.; Li, R.; Liu, Y.; Du, Y.; Liu, Y.; Sun, M.; Zhao, X.; Zhao, Q. Robotic Intracellular Pressure Measurement Using Micropipette Electrode. *Sensors* **2023**, 23, 4973. https://doi.org/10.3390/s23104973

Academic Editor: Yanling Tian

Received: 15 April 2023
Revised: 18 May 2023
Accepted: 20 May 2023
Published: 22 May 2023

1. Introduction

Cells are able to maintain positive intracellular pressure compared to the extracellular environment. Intracellular pressure is an important component of the intracellular environment, which ensures the normal physiological function of cells. It has been found to play an important role in cell division [1,2], differentiation [3,4], migration [5], diseases [6,7], and embryonic tissue development [8,9]. Moreover, the existence of intracellular pressure affects the results of cell micromanipulation. For example, during cell microinjection, the positive intracellular pressure increases the resistance to the material delivery, resulting in lower deposition volume [10,11]. Furthermore, intracellular pressure supports cell swelling, making the cell membrane more resistant to external pressure. This can affect the measurement results of cell elasticity. Therefore, intracellular pressure measurement is beneficial for revealing the mechanism of cells' physiological activities and improving the accuracy of cell micromanipulation.

At present, some methods have been developed to measure the intracellular pressure of different types of cells, according to references. For example, according to Laplace's law, intracellular pressure has been estimated indirectly from the cortical tension of cells [12,13]. However, assuming the cell to be a liquid ball makes that method applicable to cells with thin membranes. For oocytes or embryos with thick zona pellucid (ZP), the measurement errors of that method may be too large to be applicable. Before and after the release of intracellular pressure caused by cutting off part of the ZP [14], the variations in the obtained poking force–cell deformation curves have been utilized to measure the intracellular pressure. Although this method is applicable for oocytes or embryos with thick ZPs, it uses laser cutting to release intracellular pressure and microforce sensors to obtain the poking force–cell deformation curves, which requires multiple, expensive pieces of equipment, significantly limiting its application. In previous research, we proposed an intracellular pressure measurement method based on the modeled relationship between the applied pressure and the deposition volume of oil into cells [11]. Although this method can be realized on a traditional cell manipulation system setup, the oil droplets injected into the cells cause significant damage to cell viability, limiting its application. Commercial intracellular pressure measurement devices with pressure feedback have also been used to measure intracellular pressure [15–17]. Although this system can measure the intracellular pressure of a variety of cells with limited harm to cell viability, the measurement device is very expensive, thus, significantly limiting its widespread use. Therefore, a simple intracellular pressure measurement method that is applicable to traditional cell manipulation tools and causes less harm to cell viability is still highly desired.

This paper proposes a robotic intracellular pressure measurement method using a traditional micropipette electrode. First, the measured resistance of the micropipette electrode filled with KCl solution and immersed inside the culture medium is modeled to analyze its variation trend when the pressure inside the micropipette increases. Then, the concentration of KCl solution producing a "steep slope" in the electrode resistance–pressure curve is determined as the one suitable for intracellular pressure measurement. Further, the measured resistance of the micropipette electrode inside the cell is modeled to obtain the intracellular pressure. Intracellular pressure is measured by the difference between two key pressures, which correspond to injection pressures P_{I1} and P_{I2}, when the micropipette electrode resistance reaches a quasi-stable state before and after the release of intracellular pressure. Based on the above work, a robotic measurement procedure of the intracellular pressure is established based on a traditional micropipette electrode system. After two weeks of continuous practice, scholars are able to successfully measure approximately 10 cells per day using commercial equipment. An experienced expert should expect to be able to measure 50–100 cells per day [15]. The experimental results on porcine oocytes demonstrate that the proposed method can operate on cells at an average speed of 20~40 cells/day with measurement efficiency comparable to commercial measurement devices. The average repeated error of the relationship between the measured electrode resistance and the pressure inside the micropipette electrode is less than 5%, and no observable intracellular pressure leakage was found during the measurement process, which guarantees the measurement accuracy of intracellular pressure. The measured results of the porcine oocytes are in accordance with those reported in related work. Moreover, a 90% survival rate of operated oocytes was obtained after measurement, proving limited damage to cell viability.

2. Materials and Methods

2.1. System Setup

We conducted the intracellular pressure measurement experiment on the micropipette electrode system developed in our laboratory. The system setting is shown in Figure 1. The 35 × 10 mm culture dish was placed on a fixed platform on the vibration isolation platform, in which the cell operating fluid and experimental samples were placed. A standard upright microscope (Eclipse FN1, Nikon, Tokyo, Japan) capable of movement in the XY plane with a repeatability of ±0.1 μm over an area of 2 × 2 cm^2 was used to observe cells in the intracellular pressure measurement experiment. The motorized focusing device on the microscope was used to automatically focus the cells with a repeatability of ±0.1 μm in the vertical direction. On the left arm was an X-Y-Z micromanipulator model MP285 (stroke space 2 × 2 × 2 cm^3, maximum speed 1 mm/s, repeatability ±0.1 μm) for mounting micropipette electrodes. The right arm was a scientific model X-Y-Z micromanipulator (stroke space 2.5 × 2.5 × 2.5 cm^3) for mounting the holding pipette (HP). An in-house developed pneumatic chamber [18] was used to provide pressure inside the micropipette. A charge-coupled device (CCD) camera (IR-2000, DAGE-MTI, Michigan City, IN, USA) was mounted on the microscope to acquire images of the cells at 60 frames per second. A host computer was used for microscopic image processing, electrical signal acquisition, suction pressure control, and motion control of the microscope and manipulator. The entire robot system was covered with an electromagnetic shield to isolate electrical interference from the external environment. A human–machine interface based on Qt programming (see Figure 2) was developed to provide visual feedback and display information about the process of measuring intracellular pressure, suction pressure, electrical signals, output pressure, etc. This interface allows the operator to update the pressure applied inside the micropipette and the distance the micropipette electrode moves at each step. The electrical signal detected by the micropipette electrode was amplified using an amplifier (MultiClamp 700B, Molecular Devices, San Jose, CA, USA), then converted into a digital signal, and finally transmitted to the host computer.

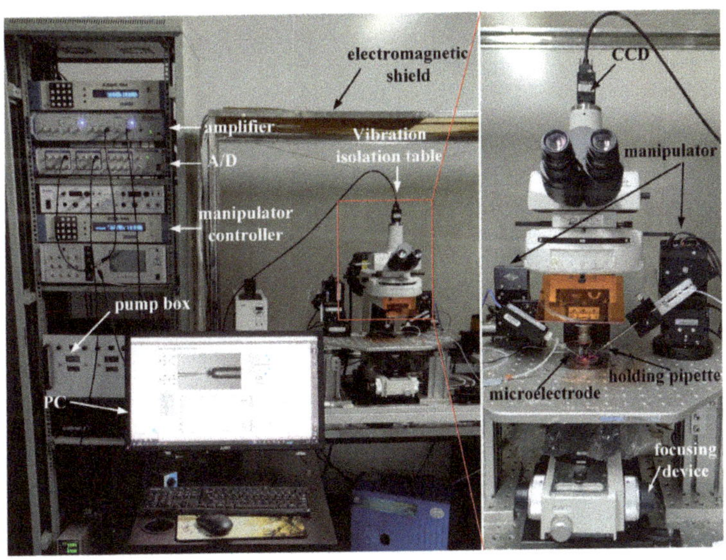

Figure 1. System setup.

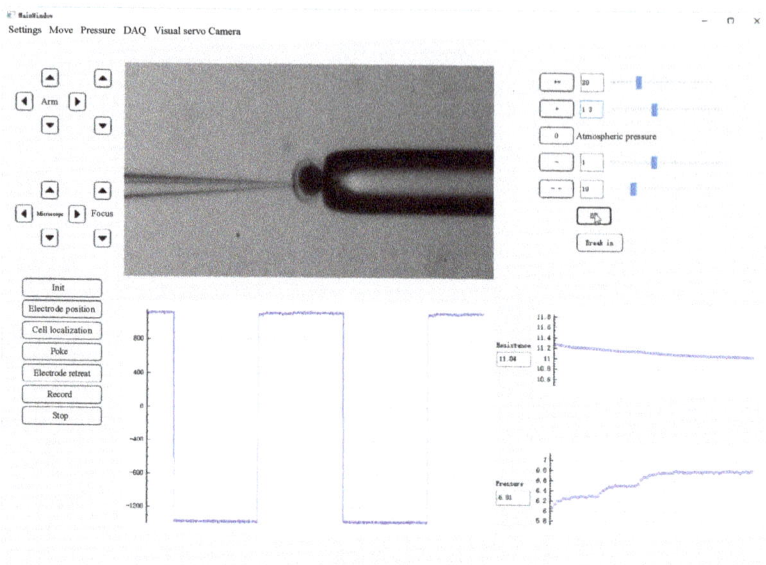

Figure 2. Human–machine interface.

2.2. *Fabrication of Micropipette Electrode and Oocyte Preparation for Intracellular Pressure Measurement*

The micropipette electrode used in the intracellular pressure measurement experiment was made of a glass tube (BF150-117-10, Sutter, Sacramento, CA, USA) with an inner diameter of 1.17 mm and an outer diameter of 1.5 mm. The glass tube was pulled with a micropuller (P97, Sutter) to form a tapered tip with a diameter of ~2 μm (see Figure 3a). If that micropipette were mounted directly on the robot arm, the micropipette electrode would penetrate the cell at an oblique 30°. This is because the electrode holder on the robot is installed at an inclination of 30° (see Figure 3b) in consideration of the use limitation of space, the stroke of the robot, etc. Therefore, to reduce the torsional effect of micropipette penetration on cells, we used a micropipette forging instrument to bend the tip of the micropipette by 30° (see Figure 3c). In this way, the tip can be horizontal after being installed on the robot. In the same way, we also bend the holding micropipette by the same degree before mounting it on the right arm. In this way, the axes of the cell, micropipette electrode, and holding pipette are in a horizontal state during the experiment.

Then, the micropipette was backfilled with 20 μL electrode liquid. Before the measurement, the cell culture medium was replaced by extracellular fluid (g/L: HEPES 0.6911, $C_{16}H_{18}N_2O_4S$ 0.05, $NaHCO_3$ 2.2504, NaCl 1.7532, $C_{21}H_{39}N_7O_{12}$ 0.06, M199 9.5, pH 7.0~7.2) for electrical signal recording. A silver electrode wire with a diameter of 0.2 mm was inserted into the micropipette to form the micropipette electrode. One section of the micropipette electrode was immersed in the extracellular fluid, and the ground wire connected at the other end was directly placed in the extracellular fluid, thus forming a complete circuit.

The oocytes we used in the experiment were obtained using the following methods. We collected pig ovaries from the local slaughterhouse and used a sterile 10mL syringe with an 18-gauge needle to aspirate ovarian follicles with a diameter of 3–6 mm to obtain cumulus–oocyte complexes (COCs). At this time, there are many tissue impurities in the COCs liquid. To obtain clean COCs, we poured them into Tyrode's lactate (TL)–Hepes-PVA (polyvinyl alcohol, 0.1%) for washing. After washing, we waited for their precipitation for 20 min and then poured out the supernatant to wash and precipitate again, repeating three times. Then, 20~40 COCs were picked up and transferred to 100 μL mature medium (TCM-199 supplemented with 15% FBS, 10 ng/mL EGF, 10% porcine follicular fluid, 10 IU/mL of eCG, 5 IU/mL of hCG, 0.8 mM L-glutamine, and 0.05 mg/mL gentamicin) for 42 h at

38.5 °C, 5% CO_2, and saturated humidity. At the MII stage, we selected target oocytes that had a clear perivitelline space, integrated cell membrane, and a visible polar body (PB1) (Figure 3d).

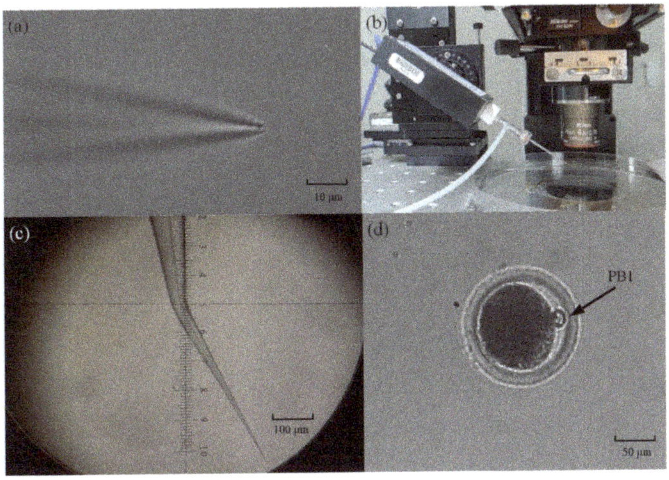

Figure 3. Fabrication of micropipette electrode and oocyte preparation: (**a**) Imaging of a drawn micropipette under a 40× microscope; (**b**) The angle between the micropipette electrode and the horizontal plane is 30°; (**c**) The micropipette is bent on the forging needle instrument; (**d**) Porcine oocytes with first polar body.

2.3. Resistance Analysis for Micropipette Electrode Out of the Cell

It is well known that when the glass micropipette is filled with electrolyte solution inside and outside, and an electromotive force is applied, the glass micropipette can form a stable resistance.

The tip diameter used in our experiment was ~2 μm pipette, whose geometric structure is similar to an infinite cone (Figure 4). We chose KCl solution as the internal solution of the electrode and assumed that KCl diffusion, just like in isotropic homogeneous media, increases the KCl flux due to a large amount of flow:

$$\frac{\partial C}{\partial t} = D\frac{\partial^2 C}{\partial r^2} + \frac{2D}{r}\frac{\partial C}{\partial r} - v(r)\frac{\partial C}{\partial r} \quad (1)$$

where C = KCl concentration (moles/liter), D = KCl diffusion coefficient (cm^2/s), r = distance from origin of cone (cm), and $v(r)$ = fluid velocity at r (cm/s).

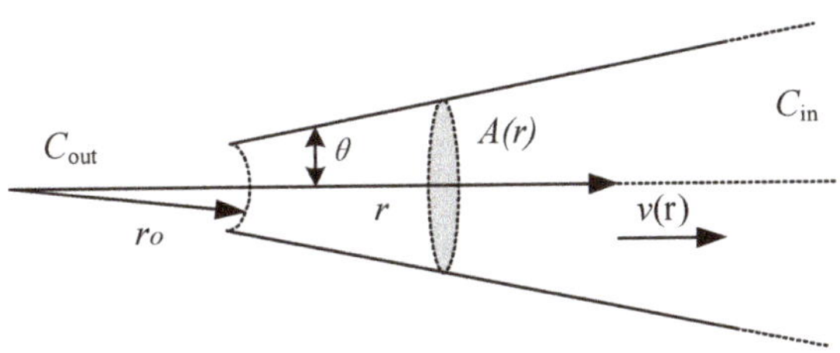

Figure 4. Schematic diagram of micropipette structure.

It is assumed that the fluid velocity is the same on any given cross-section ("plug" flow), and the velocity passes through the origin of the cone (which allows a one-dimensional solution in a spherical coordinate system). For a given flow, the speed of each cross-section will be inversely proportional to the square of the radius of that cross-section (Equation (2))

$$v(r) = (k/r^2)(P_{out} - P_{in}) \tag{2}$$

where k = a constant containing geometrical terms and pipet hydraulic resistance, P_{in} = pressure applied to the inside of the pipet (mm Hg), and P_{out} = outside pressure (mm Hg).

We can calculate the total pipette resistance by integrating the conductivity on the length of the pipette.

$$R = \int_{r_o}^{\infty} \frac{1}{K_{sp}A(r)} dr \tag{3}$$

where R = total pipet resistance (Ω), r_o = distance of the tip from the origin of the cone (cm), and $A(r)$ = cross-sectional area at r (cm^2).

The electrode resistance is affected by the solution concentration. When the solution concentration decreases, the measured electrode resistance increases. When the solution concentration (less than the solubility of the solution itself) increases, the measured electrode resistance decreases.

When the micropipette electrode is out of the cell, the measurement micropipette electrode circuit is shown in Figure 5a. The micropipette electrode measurement resistance R_M is composed of the silver wire electrode resistance R_E, the solution resistance R_L, and the resistance R_I caused by the concentration gradient field of the electrode solution and the extracellular solution near the micropipette electrode opening, namely

$$R_M = R_E + R_L + R_I \tag{4}$$

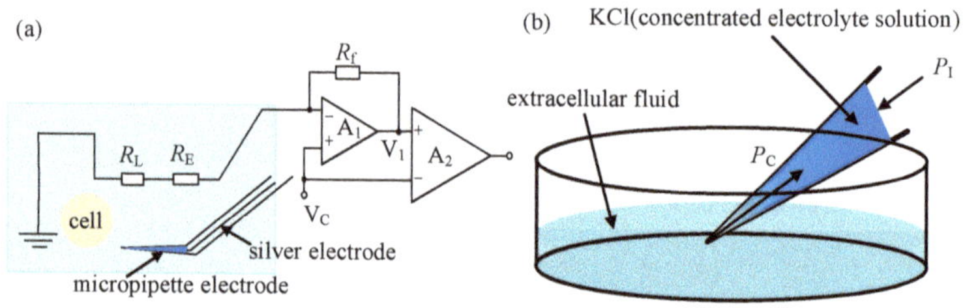

Figure 5. Resistance analysis for micropipette electrode out of the cell: (**a**) The measurement micropipette electrode circuit; (**b**) Ion concentration distribution in micropipette electrode.

Among them, R_E and R_L can be considered to be constant when the volume of the liquid enters or comes out of the micropipette is negligible in comparison to the whole volume of liquid inside the micropipette and the environment. For R_I in Equation (4), the ion concentration gradient field corresponding to R_I outside of the cell is formed by the combined action of the following two factors: capillary pressure P_C between the tube wall and the liquid in the electrode and the injection pressure P_I in the microtubule, as shown in Figure 5b. When P_I is small, capillary pressure P_C will press extracellular fluid (dilute electrolyte solution) into the electrode to form a stable ion concentration gradient and P_I value. When the P_I increases, the ion concentration distribution at the tip will shift to the opening, resulting in an increase in the concentration of KCl at the tip, which will lead to a decrease in the resistance of the electrode. When P_I is larger enough, the concentration gradient area will be pushed out of the micropipette opening. In that case, R_M will reach a stable state with a further increase in P_I if the deposition volume of the solution into the environment is negligible in comparison to the whole volume of solution inside the micropipette.

In summary, a "downslope" section of measurement resistance of the micropipette can be obtained with an increase in P_I. When the micropipette enters intracellular space, the intracellular pressure leads to a shift of concentration gradient inside the micropipette and subsequently changes the measured resistance of the micropipette electrode. Thus, the variation of the measured electrode resistance can be utilized to measure the intracellular pressure.

2.4. Prepare the Required Electrode Internal Liquid and Measure the Working Range of the Probe Micropipette Electrode

As aforementioned, the variation in the measured resistance of the micropipette electrode can be utilized to measure intracellular pressure. In the experiment, we chose KCl solution as the electrode solution. To guarantee measurement accuracy, an appropriate concentration of the solution producing a steep downslope in the R_M-P_I relationship is preferred because the R_M is more sensitive to the variation of P_I in that case, which improves the measurement accuracy of P_I. An increase in KCl concentration increases the height of the downslope, according to the analysis in Section 2.3. However, a too-large KCl concentration will increase the crystallization of KCl molecules and increase the blocking issues of the micropipette opening. Thus, an appropriate KCl concentration needs to be determined through tests.

The concentration gap of the KCl solution starts to increase from 0.1 mol (pure water) to 3 mol with an interval of 0.1 mol until the experiment is stopped due to the crystal blocking of the nozzle due to the excessive concentration of the solution. According to testing results, we found that the decreasing speed of R_M increases in the slope area as the concentration of KCl solution increases at the beginning. After the concentration is larger than 1 mol, the blocking incidences start to occur, and its occurrence rate grows as the concentration increases. Therefore, in the following experiment, we chose KCl with a concentration of 1 mol/L as the electrode internal solution. Figure 6 shows the obtained R_M-P_I curve. It can be found that a steep downslope section, which is the ideal working section for intracellular pressure measurement, exists in the obtained R_M-P_I curve.

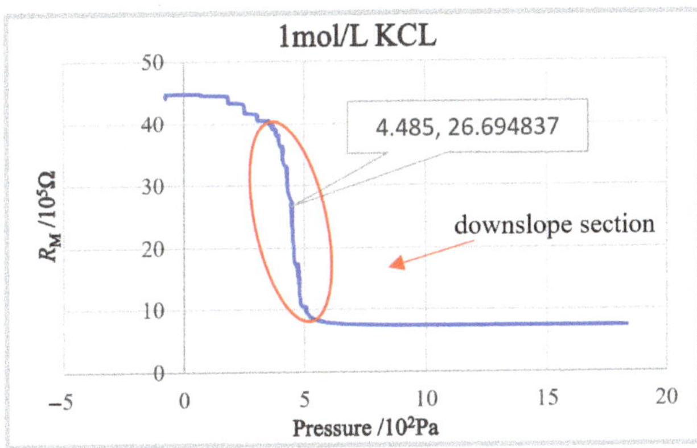

Figure 6. The R_M-P_I curve with a downslope section.

2.5. Intracellular Pressure Measurement Based on Electrode Resistance Model

To measure the intracellular pressure, the measurement resistance of the micropipette electrode is modeled, as shown in Figure 7a. When the micropipette electrode enters the cell, the zona pellucida resistance R_{ZP} and the cytoplasmic resistance R_C also become part of the measurement resistance R_M, namely

$$R_M = R_E + R_L + R_I + R_C + R_{ZP} \tag{5}$$

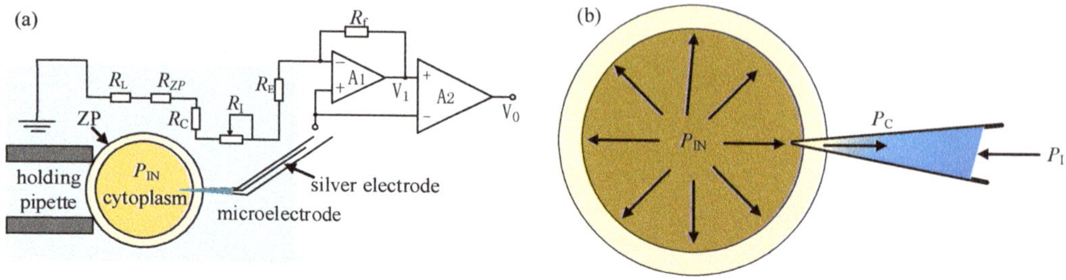

Figure 7. Electrode resistance model for measuring intracellular pressure: (**a**) The resistance model of micropipette electrode; (**b**) The ion concentration distribution in the micropipette electrode after cell penetration.

Among them, R_E, R_L, and R_{ZP} can be regarded as constants in the experiment. The cytoplasmic resistance R_C can be regarded as constant on the premise that the micropipette electrode is sealed with the ZP and the amount of injected material is far less than the cytoplasmic volume during the puncture process. After the micropipette electrode is inserted into the cell before the intracellular pressure is released, the ion concentration gradient field corresponding to R_I is formed by the joint action of the intracellular pressure P_{In}, the injection pressure P_I, and the capillary pressure P_C, as shown in Figure 7b. Adjusting the P_I, R_I, and R_M will also change accordingly. When the applied P_I remains unchanged and the change amplitude of R_M within a certain period of time does not exceed 3% of the current value, it is considered to have reached a quasi-stable state. Record the P_{I1} at this time and retreat the micropipette electrode. Then, the micropipette electrode enters intracellular space again after the release of intracellular pressure. The ion concentration gradient field corresponding to R_I is formed again by injection pressure P_I and capillary pressure P_C. Then, a new stable state of R_M similar to the former one before intracellular pressure release can be achieved through adjustment of P_I again. The difference between the two injection pressures P_I forming the stable states of R_M before and after the release of intracellular pressure separately can be considered as the released intracellular pressure value.

2.6. Robotic Measurement Procedure of Intracellular Pressure Measurement

A robotic measurement procedure of intracellular pressure was developed based on the above work. Figure 8 summarizes the robotic measurement procedure.

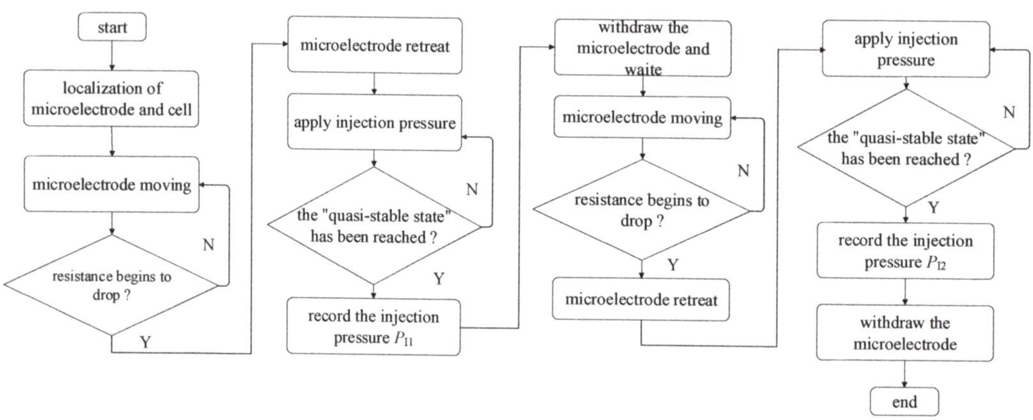

Figure 8. Flow chart for robotic measurement of intracellular pressure.

Before measurement, the micropipette electrode and the holding micropipette were moved into the field of vision manually. Then, one oocyte was put into the field of view. The system automatically focuses and 3D localizes the micropipette electrode, holding micropipette, and the target oocyte [19,20], corresponding to the frame "localization of microelectrode and cell" in Figure 8. Then, the system controlled the holding micropipette to approach the target oocyte and immobilize it with aspiration pressure [21].

Before cell membrane (zona pellucid (ZP) for porcine oocyte in this paper) penetration, the injection pressure inside the micropipette electrode was adjusted to make the resistance of the micropipette electrode within the working range (downslope section). After that, the micropipette electrode was controlled to penetrate the cell membrane along the central axis of the holding micropipette, and the system automatically detected the resistance of the micropipette electrode, corresponding to the frame "microelectrode moving" in Figure 8. Once the micropipette electrode contacts the cell, the detected resistance will start to rise, and the resistance will continue to rise as the micropipette electrode advances. When the detected micropipette electrode resistance begins to decrease, which means the micropipette has entered the intracellular space, the system assumes that the cell membrane has been punctured, corresponding to the frame "resistance begins to drop?" in Figure 8. To reduce the squeezing of micropipette electrode cells, which usually generates a false larger intracellular pressure, the system controls the micropipette electrode to retreat until the cytoplasm contour recovers its sphere shape, which means the deformation resulting from cell membrane penetration fully recovers, corresponding to the frame "microelectrode retreat" in Figure 8. The contour detection method for cytoplasm has been reported in our previous research [22].

After the cell membrane penetrates, the system increases P_{I}, and R_{M} decreases accordingly, corresponding to the frame "apply injection pressure" in Figure 8. This is because a 1 mol/L KCl solution is a concentrated electrolyte solution compared to the intracellular fluid. When the applied P_{I} remains unchanged and the amplitude of R_{M} change within 5 s does not exceed 3% of the current value, it is considered to have reached a quasi-stable state, corresponding to the frame "the quasi-stable state has been reached?" in Figure 8. P_{I1} was recorded at this time, and the micropipette electrode was withdrawn from the cell along the axial direction of the holding micropipette and waited for 60 s to release the intracellular pressure, corresponding to frames "record the injection pressure P_{I1}" and "withdraw the microelectrode and wait" in Figure 8. Then, the system controlled the holding pressure of the holding micropipette to first spit and then hold the cell again. In this way, the cell rotated by some degrees relative to the original position. Then, the micropipette electrode entered the cell again along the axis direction, retreated to let the cell recover its shape, increased the injection pressure again to reach the quasi-stable state, and recorded the required injection pressure P_{I2} at this time. The intracellular pressure of cells is then determined by the difference in injection pressure before and after the release of intracellular pressure. The corresponding P_{I2} is considered valid only when the difference in resistance values between two quasi-stable states does not exceed 100 Ω, which means that the measurement results can be adopted. We have established a flowchart for the robotic measurement of intracellular pressure (see Figure 8).

The reason why we make the cell rotate relative to the original position before entering the cell again is given as follows. The micropipette electrode usually releases a small amount of electrode fluid (concentrated electrolyte solution) around the point it stays after the first puncture. Due to the poor diffusion of the KCl solution inside the cell in comparison to that outside the cell, there may still be some KCL solution around the previous point. If the micropipette tip stays at the same position after the second entry, the existence of the left KCl solution may weaken the concentration gradient and the variation of R_{M} along P_{I}, increasing the difficulty of obtaining P_{I2}. Therefore, cell rotation before the second entry is required to make the position it stays inside the cell different from the first one. This ensures that the micropipette electrode tip is filled with KCl solution on the inner side and pure intracellular liquid on the outside during the two penetration processes of the cell.

3. Results

3.1. Intracellular Pressure Measurement Results

We used porcine oocytes in the MII period as experimental samples. The extracellular fluid was the cell operating fluid (H199) that could keep the oocytes active in vitro. The micropipette electrode was filled with 1 mol/L KCl solution. All experimental animal protocols were approved by the Institutional Animal 129 Care and Use Committee of Nankai University and carried out following the national 130 ethical guidelines for laboratory animals (permission number SYKX 2019-0001).

Figure 9a shows the first penetration of the micropipette electrode into the cell, while Figure 9b shows the second penetration after the release of intracellular pressure. Figure 9c–f shows the detection resistance and injection pressure in our experiment, respectively. Figure 9c,d shows data sampling after the first puncture. Figure 9e,f shows data sampling after the intracellular pressure is released and then inserted again. From Figure 9, we can see that after the micropipette electrode is inserted into the cell for the first time and the detection resistance reaches the quasi-stable state, the corresponding injection pressure P_{I1} = 1288 Pa. The corresponding injection pressure P_{I2} = 681 Pa after the second insertion and the detection resistance is stable. Based on the above research, we can calculate the intracellular pressure of this oocyte as $P_{in} = P_{I1} - P_{I2}$ = 607 Pa. According to statistics, the intracellular pressure of porcine oocytes we measured is 200~700 Pa with an average value of 500 $\pm$ 50 Pa (n = 20), which is basically consistent with the numerical range reported in the relevant literature [13,14]. This consistency proves the effectiveness and feasibility of our measurement method.

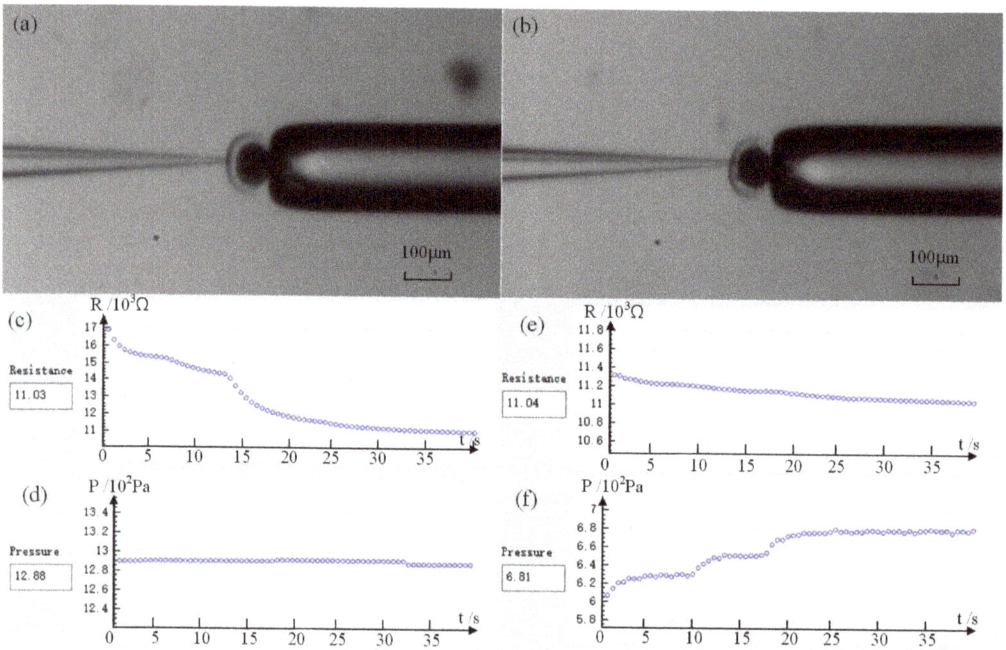

Figure 9. Intracellular pressure measurement results: (**a**) Micropipette electrode penetrates cell before releasing intracellular pressure; (**b**) Micropipette electrode penetrates cells after the release of intracellular pressure; (**c**,**d**) Resistance value and corresponding pressure in the first quasi-stable state; (**e**,**f**) Resistance value and corresponding pressure in the second quasi-stable state.

Finally, after one week of practice, the average measurement speed of the proposed method is 8 $\pm$ 2 cells/h. During the experiment, it is usually necessary to change a new micropipette after measuring a cell. This is because multiple punctures into the cells will cause some cytoplasm to adhere around the micropipette nozzle, which will lead to unstable detection resistance. Actually, the replacement and preparation of new micropipette costs most of the operation time for each cell (about 5 min). The measurement success rate is 50% (n/m) at present. With this measurement efficiency, the system is expected to measure about 20~40 cells a day, totally comparable to that of the method using commercial devices [15], which are usually able to measure 10 cells per day after two weeks of practice. We compare the operating time of our method with that of using commercial equipment methods in Table 1. Our method is fast enough to measure intracellular pressure compared to previous methods.

Table 1. Comparison between our method and previous work.

Comparison	Our Method	Commercial Equipment	Oil Injection Method
number of cells operated	20~40 cells/day	10 cells/day	20 cells/day
practice time	1 week	2 weeks	/
cell survival rate	90%	/	0%

3.2. *Leakage Test for Intracellular Pressure during the Measurement Process*

During the measurement process of intracellular pressure, the ZP and outside wall of the micropipette need to be well sealed to avoid leakage of intracellular pressure. Otherwise, the leakage of intracellular pressure leads to a falsely smaller value of measured intracellular pressure than its original one. In this part, the fluorescent dye solution was backfilled into the micropipette to detect possible leakage of intracell pressure during measurement.

First, a fluorescent dye (HTPS, 5 mM) solution was backfilled into the micropipette before measurement. After ZP penetration, a positive pressure pulse was exerted inside the micropipette to deposit about 2 pL volume of solution, only about 0.25% of the total volume of the oocyte calibrated by the method mentioned in previous research [11]. Then, the field of view switches to the fluorescent field to check the diffusion scope of the fluorescent dye. If any fluorescent dye diffuses outside of the cell through the wound, it means that the seal between the micropipette and the ZP has failed, and leakage of intracellular pressure occurs during measurement. According to the experimental results on 10 oocytes, we found no leakage of intracellular pressure during the measurement process (see Figure 10), which guarantees the measurement accuracy of the intracellular pressure using our method.

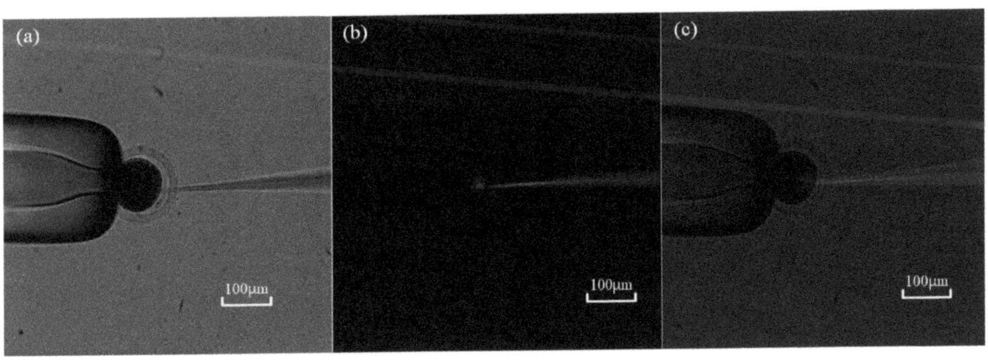

Figure 10. Leakage testing of intracellular pressure: (**a**) Micropipette electrode penetration into cells under bright field; (**b**) Observation of fluorescence diffusion within cells under fluorescence field; (**c**) Synthesis of two images under bright and fluorescent fields.

3.3. *Repeatability Validation Experiment*

Unfortunately, as an invasive measurement method, our method is not able to measure the intracellular pressure of one cell more than once because the intracellular pressure leaks from the wound after the first measurement. Thus, the repeatability of our measurement method cannot be evaluated through comparison among the obtained intracellular pressure multiple times for the same cell. However, as a measurement method relying on the R_M-P_I relationship, the repeatability of our measurement method can be indirectly evaluated through the repeatability of the obtained R_M-P_I curves. Due to the aforementioned poor diffusion rate of the released KCl solution inside the cell, the local concentration gradient at the opening of the micropipette and the obtained R_M-P_I curves can be easily varied because of the KCl solution released from the micropipette during the test. Thus, the test was conducted in H199 fluid outside the cell, where the diffusion of the KCl solution is much faster. Further, compared to the volume of KCl injected by the micropipette electrode, the volume of H199 fluid can be assumed to be infinite. Thus, the environment can be considered stable in multiple experiments.

After the test starts, the micropipette electrode is immersed in the H199 solution. Then, injection pressure inside micropipette P_I is back-and-forth adjusted to record the corresponding measurement resistance, R_M, repeated three times. The obtained R_M-P_I curves are summarized in Figure 11. We can see that when the recorded pressure is greater than 300 Pa, the obtained R_M-P_I curves almost coincide with each other. When the resistance value is the same, the pressure value error is less than 47 Pa. The average error is 22 Pa, which is only 4.4% of the average measured intracellular pressure. It is worth noting that the consistency of the three curves reduces when P_I is less than 300 Pa. This is mainly because the openness of the valve to adjust the P_I is almost closed when the output pressure is less than 300 Pa, which easily produces an unstable air flow, resulting in unsteady R_M. Fortunately, the diameter used in this article is ~2 μm. To keep the resistance value of the micropipette electrode within the working range (downslope section) and prevent the tip from being blocked in the process of penetrating the cell, the system will give the micropipette electrode a positive pressure greater than 300 Pa before the penetration. Therefore, R_M disturbances below 300 Pa have no significant impact on our measurement results. The good repeatability of the R_M-P_I curve obtained ensures the repeatability of the proposed intracellular pressure measurement method.

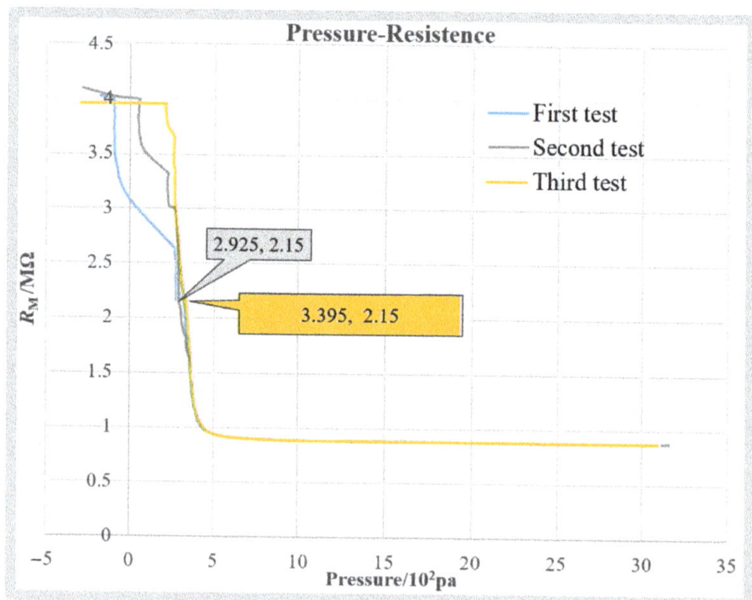

Figure 11. Repeatability validation experiments.

3.4. Cell Survival Rate Testing Results after Measurement

The survival rate of the measured oocyte was tested after intracellular pressure measurement to evaluate the harm caused by the proposed method to the oocyte viability. The operated oocyte is considered to survive if a clear and integrated cytoplasm membrane exists (see Figure 12a), which means the wound has healed after 2 h of culture. If part of the cytoplasm membrane still breaks at that time (see Figure 12b), which means the wound is still open, the viability of the oocyte is considered to be damaged, and the oocyte may be dead. According to the observation results, only two oocytes with a broken cytoplasm membrane were found from the twenty oocytes with successful measurement. The above results demonstrate that the survival rate of the oocyte measured by the proposed method is up to 90% (18/20), proving limited harm to cell viability. This is very important for biological applications requiring alive oocytes or embryos, such as oocyte or embryo microinjections. For these applications, the in situ measurement of intracellular pressure can be achieved using our method before operations.

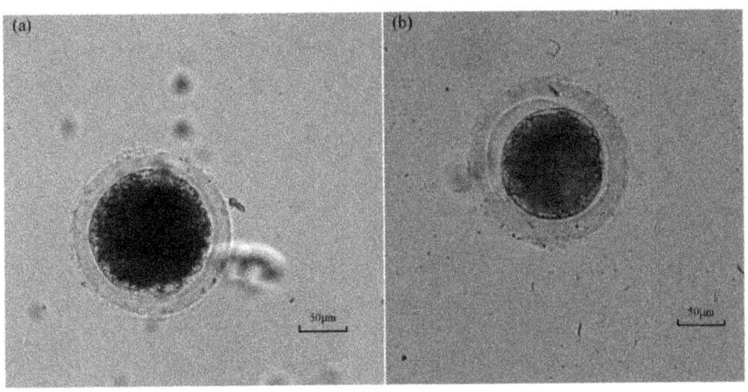

Figure 12. Oocyte images after 2 h of culturing after the measurement: (**a**) The oocyte with part of a broken membrane which may be a dead oocyte. (**b**) Alive oocyte with an integrated cytoplasm membrane.

The use of the oil injection method to measure intracellular pressure is also an invasive measurement method [11]. However, as is well known, mineral oil is a harmful substance to cells, and cells measured using this method are almost unable to maintain activity. Table 1 compares our method with the oil injection method in terms of cell survival rate after measuring intracellular pressure. Our method causes less damage to cells compared to the oil injection method.

4. Discussion

Our method can be conducted using a common microoperation system with the measurement ability of the micropipette electrode resistance. Because it does not involve expensive specific commercial measurement devices, which usually cost USD 80,000–90,000 [23], the proposed system is expected to have more widespread applications. Through comparing the required P_I making R_M reach a quasi-stable state before and after the release of the intracellular pressure, the influences of the cytoplasm and ZP on the R_M are excluded, which may improve the measurement accuracy of intracellular pressure in comparison to the method based on commercial devices [15].

Although our system is only able to measure one cell at one time at present, it is highly expected to achieve robotic measurement of batch cells in the future. We proposed a robotic enucleation method [20] for batch oocytes in the previous study. In that method, the system automatically scanned the entire area of oocytes and established a global map containing all the target oocytes before operation. With the global localization of the target oocytes, the system was able to pick up the target cell with the holding micropipette, move the cell to another place with movement of the microstage, and then operate it with the injection micropipette one by one. Further, the pancreatin solution has been proven to be able to clear the inner surface of the micropipette and remove the adherent cell materials on it [24]. With a droplet of pancreatin solution to wash the micropipette after measurement on each cell, the replacement of the micropipette can be saved. Similar functions can be integrated into our system for the robotic measurement of batch cells in the future. After that, the time cost in search of each cell and replacement of the micropipette can be saved, which will significantly accelerate the operation speed of our method.

At present, the success rate of the proposed system is about 50%. The two main reasons for the measurement failures are the blockage of the micropipette electrode opening and failures of quasi-stable state formation after ZP penetration. Apparently, the increase in the diameter of the micropipette electrode and the decrease in the concentration of KCl concentration are both able to reduce the blockage of the micropipette opening. However, a micropipette tip with a larger opening increases the cut size generated during penetration and subsequently increases the leakage incidences of the intracellular pressure. Moreover, the decrease in KCl concentration may further reduce the concentration gradient at the micropipette opening after ZP penetration, which finally reduces the formation success rate of the quasi-stable state of R_M. Sharpening the tip with a tilt opening may be a better way to reduce the blockage in the future, as it can increase the opening area of the micropipette and does not enlarge the wound. Increasing the diffusion speed of the solution may increase the formation success rate of the quasi-stable state. Thus, a compound solution with a diffusion speed higher than KCl is expected to replace the KCl solution in the future for the improvement of the formation success rate of the quasi-stable state of R_M. In addition, after completing the experiment, some KCl solution will be left in the cell, which may affect the osmotic pressure of the cell. Therefore, we speculate that the left KCL has an impact on cell viability. However, from our conclusion in Section 3.4, we can see that the cell survival rate reaches 90%, so we can infer that the impact of left KCL on the cell survival rate is very small.

Additionally, it has been reported that the speed at which the micropipette penetrates cells can have an impact on cell deformation and damage, and optimizing the penetration speed can reduce the damage caused by micropipette penetration to cells [25]. This inspires the potential improvement of the work in this paper. Currently, in our experiment, the micropipette electrode penetration into cells is at a constant speed of 10 μm/s. In the future, we can control the penetration speed to reduce cell damage.

5. Conclusions

In this paper, a method of intracellular pressure measurement based on a micropipette electrode circuit model is proposed. We first introduced the fabrication method of the micropipette electrode used in the experiment. Secondly, according to the change principle of electrode detection resistance, we established the circuit model of the micropipette electrode outside the cell and carried out the extracellular injection experiment to determine the appropriate concentration of the electrode liquid used in the experiment. Then, we established the circuit model of the micropipette electrode inside the cell. Finally, a robotic measurement procedure was established based on the above work. We carried out our intracellular pressure measurement experiment on the patch clamp system in our laboratory and obtained the intracellular pressure of porcine oocytes (200~700 Pa), which is basically consistent with the value reported in the relevant literature. The effectiveness and feasibility of our proposed method were validated through experiments. Although our experiment was implemented on a patch clamp system, this method is easy to generalize to conventional micropipette electrode resistance measurement systems, so it does not require expensive systems and helps achieve in situ detection.

Author Contributions: Q.Z., X.Z. and M.L. conceived the idea for the study. Q.Z., X.Z., M.L. and J.Q. designed the experiments. M.L., R.L. and J.Q. performed the experiments. Q.Z. and X.Z. supervised the project. Q.Z. and M.L. wrote the manuscript. Y.L. (Yuzhu Liu), Y.D., Y.L. (Yaowei Liu) and M.S. checked typesetting and English grammar. All authors have read and agreed to the published version of the manuscript.

Funding: This work was jointly supported by the National Natural Science Foundation of China (62027812, 62273186) and the Beijing Advanced Innovation Center for Intelligent Robots and Systems under Grant 2019IRS05.

Institutional Review Board Statement: All the procedures were approved by the Animal Care and Use Committee of Tianjin Animal Science and Veterinary Research Institute and were performed in accordance with the NIH Guide for the Care and Use of Laboratory Animals (No. 8023, revised in 1996).

Informed Consent Statement: Not applicable.

Data Availability Statement: The raw data supporting the conclusions of this article will be made available by the authors, without undue reservation.

Conflicts of Interest: The authors declare no conflict of interest.

References

1. Gudipaty, S.A.; Lindblom, J.; Loftus, P.D.; Redd, M.J.; Edes, K.; Davey, C.F.; Krishnegowda, V.; Rosenblatt, J. Mechanical stretch triggers rapid epithelial cell division through Piezo1. *Nature* **2017**, *543*, 118–121. [CrossRef] [PubMed]
2. Tang, Z.; Hu, Y.; Wang, Z.; Jiang, K.; Zhan, C.; Marshall, W.F.; Tang, N. Mechanical forces program the orientation of cell division during airway tube morphogenesis. *Dev. Cell* **2018**, *44*, 313–325. [CrossRef] [PubMed]
3. Altman, G.H.; Horan, R.L.; Martin, I.; Farhadi, J.; Stark, P.R.H.; Volloch, V.; Richmond, J.C.; Vunjak-Novakovic, G.; Kaplan, D.L. Cell differentiation by mechanical stress. *FASEB J.* **2002**, *16*, 270–272. [CrossRef]
4. Cohen, D.M.; Chen, C.S. Mechanical control of stem cell differentiation. *Regen. Med.* **2012**, *7*, 101–116.
5. Strychalski, W.; Guy, R.D. Intracellular pressure dynamics in blebbing cells. *Biophys. J.* **2016**, *110*, 1168–1179. [CrossRef] [PubMed]
6. Mogil, R.J.; Kaste, S.C.; Ferry, R.J., Jr.; Hudson, M.M.; Mulrooney, D.A.; Howell, C.R.; Partin, R.E.; Srivastava, D.K.; Robison, L.L.; Ness, K.K. Effect of low-magnitude, high-frequency mechanical stimulation on BMD among young childhood cancer survivors: A randomized clinical trial. *JAMA Oncol.* **2016**, *2*, 908–914. [CrossRef]
7. Wo, F.; Xu, R.; Shao, Y.; Zhang, Z.; Chu, M.; Shi, D.; Liu, S. A Multimodal System with Synergistic Effects of Magneto-Mechanical, Photothermal, Photodynamic and Chemo Therapies of Cancer in Graphene-Quantum Dot-Coated Hollow Magnetic Nanospheres. *Theranostics* **2016**, *6*, 485–500. [CrossRef]
8. Vining, K.H.; Mooney, D.J. Mechanical forces direct stem cell behaviour in development and regeneration. *Nat. Rev. Mol. Cell Biol.* **2017**, *28*, 728–742. [CrossRef]
9. Nguyen, M.D.; Tinney, J.P.; Ye, F.; Elnakib, A.A.; Yuan, F.; El-Baz, A.; Sethu, P.; Keller, B.B.; Giridharan, G.A. Effects of physiologic mechanical stimulation on embryonic chick cardiomyocytes using a microfluidic cardiac cell culture model. *Anal. Chem.* **2015**, *87*, 2107–2113. [CrossRef]
10. Knoblauch, M.; Hibberd, J.M.; Gray, J.C.; van Bel, A.J. A galinstan expansion femtosyringe for microinjection of eukaryotic organelles and prokaryotes. *Nat. Biotechnol.* **1999**, *17*, 906–909. [CrossRef]
11. Wang, X.; Zhao, Q.; Wang, L.; Liu, J.; Pu, H.; Xie, S.; Ru, C.; Sun, Y. Effect of cell inner pressure on deposition volume in microinjection. *Langmuir* **2018**, *34*, 10287–10292. [CrossRef]
12. Rand, R.P.; Burton, A.C. Mechanical properties of the red cell membrane: I. Membrane stiffness and intracellular pressure. *Biophys. J.* **1964**, *4*, 115–135. [CrossRef] [PubMed]
13. Chan, C.J.; Costanzo, M.; Ruiz-Herrero, T.; Mönke, G.; Petrie, R.J.; Bergert, M.; Diz-Muñoz, A.; Mahadevan, L.; Hiiragi, T. Hydraulic control of mammalian embryo size and cell fate. *Nature* **2019**, *571*, 112–116. [CrossRef] [PubMed]
14. Wang, X.; Zhang, Z.; Tao, H.; Liu, J.; Hopyan, S.; Sun, Y. Characterizing Inner Pressure and Stiffness of Trophoblast and Inner Cell Mass of Blastocysts. *Biophys. J.* **2018**, *115*, 2443–2450. [CrossRef] [PubMed]
15. Petrie, R.J.; Koo, H. Direct measurement of intracellular pressure. *Curr. Protoc. Cell Biol.* **2014**, *63*, 12.9.1–12.9.9. [CrossRef]
16. Petrie, R.J.; Koo, H.; Yamada, K.M. Generation of compartmentalized pressure by a nuclear piston governs cell motility in a 3D matrix. *Science* **2014**, *345*, 1062–1065. [CrossRef]
17. Mosaliganti, K.R.; Swinburne, I.A.; Chan, C.U.; Obholzer, N.D.; Green, A.A.; Tanksale, S.; Mahadevan, L.; Megason, S.G. Size control of the inner ear via hydraulic feedback. *Elife* **2019**, *8*, e39596. [CrossRef]
18. Stoy, W.A.; Kolb, I.; Holst, G.L.; Liew, Y.; Pala, A.; Yang, B.; Boyden, E.S.; Stanley, G.B.; Forest, C.R. Robotic navigation to subcortical neural tissue for intracellular electrophysiology in vivo. *J. Neurophysiol.* **2017**, *118*, 1141–1150. [CrossRef]
19. Zhao, Q.; Wu, M.; Cui, M.; Qin, Y.; Yu, J.; Sun, M.; Zhao, X.; Feng, X. A novel pneumatic micropipette aspiration method using a balance pressure model. *Rev. Sci. Instrum.* **2013**, *84*, 123703. [CrossRef]
20. Zhao, Q.; Cui, M.; Zhang, C.; Yu, Y.; Sun, M.; Zhao, X. Robotic enuleation for oocytes. In Proceedings of the 9th IEEE International Conference on Nano/Micro Engineered and Molecular Systems, Waikiki Beach, HI, USA, 13–16 April 2014.
21. Zhao, Q.; Sun, M.; Cui, M.; Yu, J.; Qin, Y.; Zhao, X. Robotic Cell Rotation Based on the Minimum Rotation Force. *IEEE Trans. Autom. Sci. Eng.* **2015**, *12*, 1504–1515. [CrossRef]
22. Zhao, Q.; Qiu, J.; Feng, Z.; Du, Y.; Liu, Y.; Zhao, Z.; Sun, M.; Cui, M.; Zhao, X. Robotic Label-free Precise Oocyte Enucleation for Improving Developmental Competence of Cloned Embryos. *IEEE Trans. Biomed. Eng.* **2021**, *68*, 2348–2359. [CrossRef] [PubMed]

23. 900A Micropressure Ssytem from WPI. Available online: https://www.wpiinc.com/blog/post/900a-micropressure-system-from-wpi (accessed on 1 March 2023).
24. Álvarez-Padilla, F.; Hormiga, G. A Protocol for Digesting Internal Soft Tissues and Mounting Spiders for Scanning Electron Microscopy. *J. Arachnol.* **2007**, *35*, 538–542. [CrossRef]
25. Liu, Y.; Cui, M.; Zhang, Y.; Zhao, X.; Sun, M.; Zhao, X. Oocyte Penetration Speed Optimization Based on Intracellular Strai. *Micromachines* **2022**, *13*, 309.

sensors

MDPI

Article

A Novel Active Device for Shoulder Rotation Based on Force Control

Isabel M. Alguacil-Diego [1], Alicia Cuesta-Gómez [1,*], David Pont [2], Juan Carrillo [2], Paul Espinosa [2], Miguel A. Sánchez-Urán [2] and Manuel Ferre [2]

1 Physiotherapy, Occupational Therapy, Physical Medicine and Rehabilitation Department, Universidad Rey Juan Carlos, Campus de Alcorcón, Av. de Atenas s/n, Alcorcón, 28922 Madrid, Spain; isabel.alguacil@urjc.es

2 Centre for Automation and Robotics (CAR) UPM-CSIC, Universidad Politécnica de Madrid, C/. José Gutierrez Abascal, 2, 28006 Madrid, Spain; david.pont@upm.es (D.P.); juan.carrillo.rios@gmail.com (J.C.); paul.espinosa.peralta@alumnos.upm.es (P.E.); miguelangel.sanchezuran@upm.es (M.A.S.-U.); m.ferre@upm.es (M.F.)

* Correspondence: alicia.cuesta@urjc.es

Abstract: This article describes a one-degree-of-freedom haptic device that can be applied to perform three different exercises for shoulder rehabilitation. The device is based on a force control architecture and an adaptive speed PI controller. It is a portable equipment that is easy to use for any patient, and was optimized for rehabilitating external rotation movements of the shoulder in patients in whom this was limited by muscle–skeletal injuries. The sample consisted of 12 shoulder rehabilitation sessions with different shoulder pathologies that limited their range of shoulder mobility. The mean and standard deviations of the external rotation of shoulder were $42.91 \pm 4.53°$ for the pre-intervention measurements and $53.88 \pm 4.26°$ for the post-intervention measurement. In addition, patients reported high levels of acceptance of the device. Scores on the SUS questionnaire ranged from 65 to 97.5, with an average score of 82.70 ± 9.21, indicating a high degree of acceptance. The preliminary results suggest that the use of this device and the incorporation of such equipment into rehabilitation services could be of great help for patients in their rehabilitation process and for physiotherapists in applying their therapies.

Keywords: force control; haptic; upper limb; rehabilitation; shoulder movements; medical rehabilitation

Citation: Alguacil-Diego, I.M.; Cuesta-Gómez, A.; Pont, D.; Carrillo, J.; Espinosa, P.; Sánchez-Urán, M.A.; Ferre, M. A Novel Active Device for Shoulder Rotation Based on Force Control. *Sensors* **2023**, *23*, 6158. https://doi.org/10.3390/s23136158

Academic Editors: Xin Zhao, Mingzhu Sun and Qili Zhao

Received: 23 May 2023
Revised: 27 June 2023
Accepted: 29 June 2023
Published: 5 July 2023

1. Introduction

Shoulder rotator cuff injuries include tendinopathies and partial or complete tendon ruptures, mostly related to the supraspinatus tendon [1]. Exercise is a key component of treatment for shoulder soft tissue injuries of the shoulder. Evidence recommends exercise-based interventions to reduce pain and increase shoulder function. Evidence also suggests that progressive shoulder strengthening and stretching are effective in the management of rotator cuff injuries [2]. Exercises should be introduced as early as possible. However, in clinical practice, the actual referral rate to physical therapists is low due to the lack of access to publicly funded physical therapy in health care systems [3]. Waiting for outpatient physiotherapy services has been shown to have mixed results in clinical and health system outcomes. The review by Deslauriers et al. [4] suggests the possible detrimental effects of waiting on pain, disability, quality of life, and psychological symptoms in people with shoulder disorders. There is also evidence of higher healthcare utilization and costs for patients who wait longer before physiotherapy services [4]. The effects of waiting caused by lack of access to physiotherapy services could be mitigated by the implementation of robotic devices with which more patients can be treated at the same time with a lower number of physiotherapists. Rehabilitation robots can significantly reduce the burden of therapists by providing repetitive and precise therapy to people with upper limb impairments over a long period of time [5].

Most upper extremity devices for rehabilitation available on the market are designed to improve flexion and abduction, and not external rotation, which is a fundamental movement in the performance of activities of daily living. Rotation movements are fundamental for performing activities below the horizontal and to execute coordinated movements with the hand for an individual to locate themselves in space [6]. In addition, most studies have been carried out in stroke patients [7,8]. This equipment is based on exoskeletons [9] or haptic devices [10]. The exoskeletons typically cover the arm and forearm and are linked to the trunk using flexible components. Meanwhile, haptic devices are attached to the hand to guide the movement while exerting force.

There is great interest in developing more specific equipment for rehabilitation [11]. This equipment is focused specifically on rehabilitation and solve the constraints of generic exoskeletons, particularly, the difficulties of supporting complex movements of joints such as the shoulder. The rigid exoskeletons use special mechanisms that increase its complexity. At the end, the goal is to adjust the actuated degree of freedom of the exoskeleton to the desired rotation plane of the shoulder/arm for each specific rehabilitation treatment. In some cases, textile or other flexible materials are used to avoid movement constraints.

This article describes a one-degree-of-freedom haptic device that can be applied to perform three different exercises for shoulder rehabilitation. This device implements force control behavior. This force control entails resistance to the patient's shoulder rotation that is customized according to the patient's status. Therefore, the same equipment can be adapted to the patient's evolution and all session activities can be recorded for clinician studies. A portable and compact equipment has been designed for shoulder rehabilitation treatments.

2. Materials and Methods

For the present study, a pre- and post-treatment study was performed for an intervention session using a one-degree-of-freedom haptic device to perform three different exercises for shoulder rehabilitation.

2.1. Device

The rehabilitation device consists of the following components: the actuator, transmission components, force sensor, handle, and controller. The architecture used is shown in Figure 1. The actuator is based on a DC motor; the Maxon DCX22S has been selected for this application. Its rated operation is at 38 VDC, 5.9 Nm, and 10.5 revolutions per minute (RPM). This motor generates the required force that is applied to the patient. The controller calculates the reference value according to the desired force and information provided by sensors; then, the driver (H-bridge max14870) sets the power to the motor. The controller is run on a Texas Instruments board, specifically, the LAUNCHXL-F28379D model. This board contains a 32-bit dual-core TMS320F2-8379D microcontroller, which runs at 200 MHz. The controller closes the force control loop at 1 kHz. The transmission is made up of a cable and a pulley that transmit the actuator torque to the patient; the ratio between the motor and the pulley is 100:1 mm. A force sensor is located at the end of the cable and is also linked to the handle. The force sensor is implemented by a load cell manufactured by Futek, which is used to measure the force reflected by the motor. The sensor model used in this work is LSB205 connected to the Analogue Strain Gauge Signal Conditioner IAA100; it offers a load measurement range from 0 to 11 Kg, corresponding to 0 to 5 DC volts.

The patient's hand is positioned inside the handle, which is made of a flexible textile for a comfortable grip. Figure 2 shows an image with all the components. The patients sit in a chair, and the described components are mounted in an aluminum structure that provides a comfortable workspace.

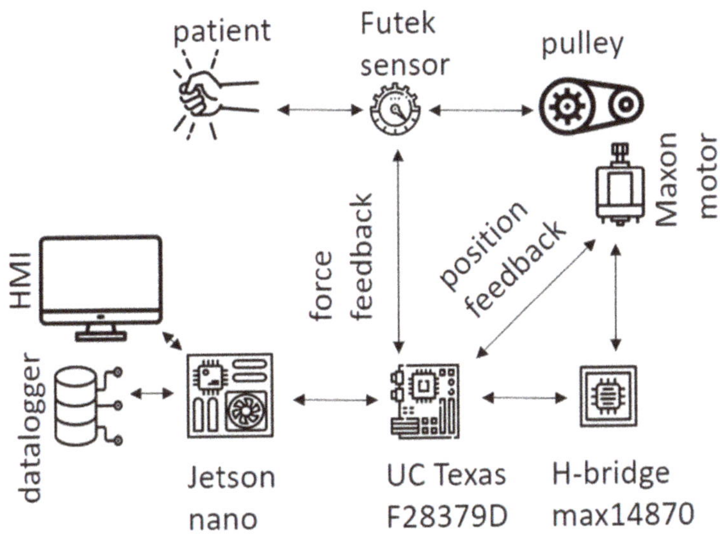

Figure 1. Proposed general architecture. The prototype consists of two electronic cards. the Jetson nano sends the setpoint parameters selected in the interface to the microcontroller (UC) Texas F28379D, and stores the values generated by the controller and the sensors. The H-bride transforms the PWM signal generated by the UC into a voltage to move the motor. The interaction between the patient and the motor is detected by the Futek load cell sensor.

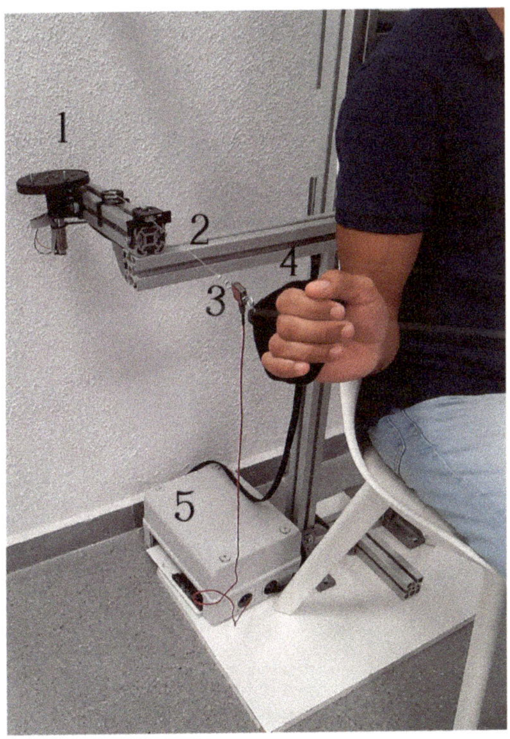

Figure 2. Main components of the rehabilitation device: actuator (1), transmission components (2), force sensor (3), handle (4), and controller (5).

2.2. Control Algorithm Design

In this work, a force controller is proposed based on the cable tension exerted by the patient and an adaptive speed PI controller. The force sensor provides a range of detection from 0 N to 111 N; therefore, the maximum configurable payload of the prototype is 11.32 Kg. The proportional gain K_p is modified as a function of the ratio between the force error and the reference; the proposed values are shown in Figure 3. The integral gain K_i was set at a value of 2. K_p and K_i were tuned experimentally.

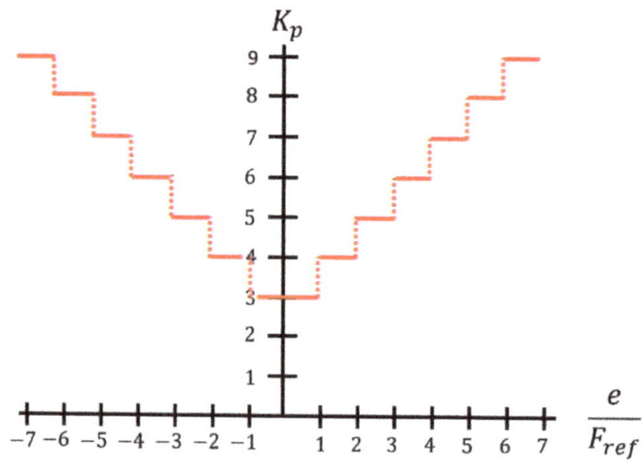

Figure 3. Value of the gain K_p according to the ratio between the force error and the reference. When the ratio increases, a higher gain is applied. The minimum value is adjusted for the recovering cable.

2.3. Controller Performance

Some trials were performed in the laboratory to properly tune the controllers. Figure 4 shows the performance of the force exerted by the motor when applied to the reference. As we can see, the information provided by the force sensor oscillates around the reference. This oscillation is due to the interaction with the patient; the different movement speed is directly transformed to an incremental variation of the reflected force. This is the usual behavior of a force control loop that maintains a reflected force.

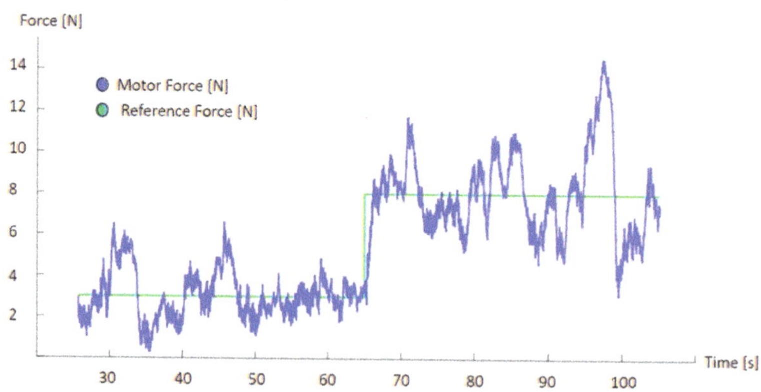

Figure 4. Evolution of the force exerted by the motor according to the force reference. Oscillations over the reference are due to the interaction with the patient to account for differences in resistance during the movement.

2.4. Participants

The inclusion criteria established for the selection of patients were as follows: upper limb musculoskeletal injury, no pathology for which rehabilitation treatment is contraindicated, and over 18 years of age. Patients who met any of the following exclusion criteria were not admitted to the study: cognitive impairment that prevented the understanding of simple commands, neurological lesion affecting the upper extremity, dermatological lesion of the upper extremity preventing the use of device material, and/or any other type of injury impeding the use of the device.

This protocol was approved by the Research Ethics Committee of the Rey Juan Carlos University Institute. The ethical principles for medical research in humans from the Declaration of Helsinki were followed. All subjects signed the informed consent form prior to participation. Trials were carried out at the Getafe Clinical Centre (Madrid, Spain).

2.5. Intervention

The clinician adjusted the controller parameters for each patient according to the predefined exercise (in this case, the shoulder rotation movement). These parameters define the shoulder rotation range and the stress of the reflected force on the patient, which is in an interval of 5 N to 15 N. Furthermore, it was verified that the pulley was in the same plane as the patient's forearm. This alignment makes movements more comfortable as the forces directly oppose shoulder rotation with no movement at the elbow.

Each session lasted 15 min and consisted of repeating a cycle of external rotation and internal rotation 30 times, returning to the starting position. These cycles were carried out with the patient sitting down, in a chair with a backrest, and with their feet on the floor. Initially, the arm was positioned close to the body with a 90° elbow flexion, and from this position, the rotation movements were performed. Figure 5 shows a patient during a session.

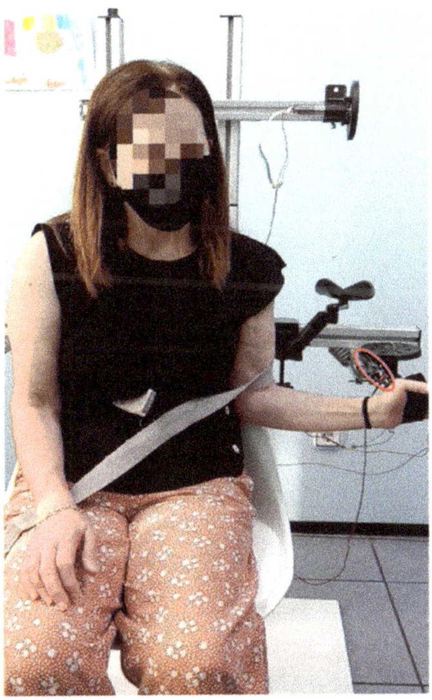

Figure 5. Patient performing shoulder rotation in the horizontal plane. The patient pulls the cable by means of a tape held in her hand. The cable is circled in red.

To perform the rotation movements, the patient started the external rotation movement from the initial position up to the maximum rotation they were able to achieve. In this half-cycle, the patient exerted a force greater than that generated by the motor. In the second half of the cycle, when the patient returned to the initial position, performing an internal rotation; this movement was in the opposite direction and the force applied by the motor was greater than that exerted by the patient. Figure 6 shows the limits of the cyclical movement. The force reflected by the actuator was fixed throughout the cycle and measured by the load cell. This load cell was located close to the patient's hand to obtain an accurate estimate of the force reflected on the patient; otherwise, the forces caused by friction might affect the information processed by the controller.

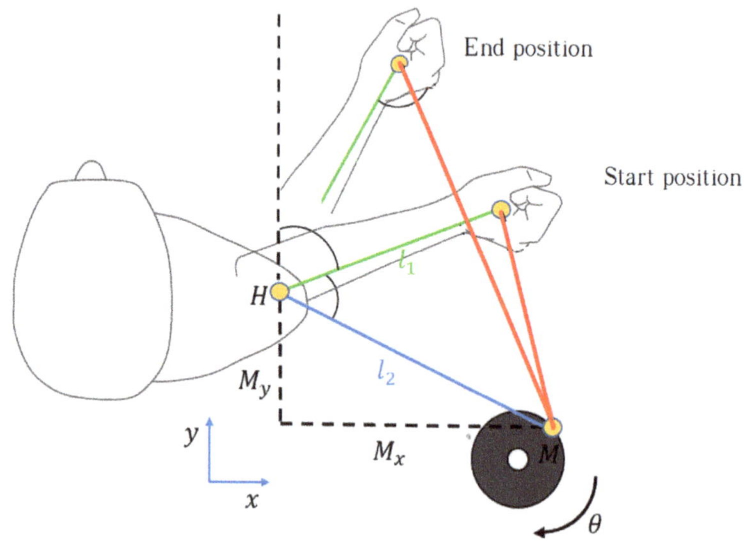

Figure 6. Range of movement for horizontal shoulder rotation. The rehabilitation session consists of the repetition of movements from the starting position to the end position. The starting position is defined by the physician and the end position is set according to the evolution of the patient.

2.6. Outcome Measures

The range of motion (ROM) of external rotation was measured using a double-armed 360° goniometer. The patients were in a supine position with the humerus abducted at 90° and the elbow flexed at 90°. Measurements were performed twice and averaged for further analysis [12].

The system usability scale (SUS) was used to assess patient satisfaction with the device. This questionnaire was developed by Brooke [13] as a usability tool and has been widely used in the evaluation of a variety of systems. The SUS provides a quick and reliable tool for measuring the usability of a device. It is a simple and short questionnaire to answer; a final score is provided with an interpretation based on a well-established reference standard. It has excellent reliability (0.85). It consists of a 10-item questionnaire with five response options, from strongly agree (score of 5) to strongly disagree (score of 1). There are five positive and five negative statements, which are presented in alternatingly. The odd-numbered questions (Q1 "I think I would use this device frequently', Q3 "I think the device was easy to use', Q5 "The functions of this device are well integrated', Q7 "I imagine that most people would learn to use this device very quickly', and Q9 "I feel safe using this device') are positive questions, and the recorded scores are the original scores subtracted by 1. The even-numbered questions (Q2 "I find this device unnecessarily complex', Q4 "I think I would need help from a person with technical knowledge to be able to use this device', Q6 "I think the device has a lack of consistency', Q8 "I find the device very difficult to use', and

Q10 "I needed to learn many things before able to use this device") are negative questions with recorded scores being subtracted from 5. Once the results for the ten questions are treated, the score of each question is added, and the result is multiplied by a factor of 2.5. The SUS score, therefore, ranges from 0 to 100, where a higher score means better usability, with a threshold of 68 to establish the usability of the device [14].

2.7. Statistical Analysis

Statistical analysis was performed using the SSPS statistical software system (SSPS Inc., Chicago, IL, USA; version 27.0). A descriptive analysis of all the variables was performed. The results were expressed as average and standard deviation and median and interquartile range. The normal distribution of the variables was verified using the Shapiro–Wilk test. The hypothesis that the variables did not have a normal distribution was accepted, given the results of this test, the verification of each variable's histograms, and the sample size. We used the Wilcoxon test, a nonparametric test for related samples. Statistical analysis was performed with a confidence level of 95%, and therefore significant values were those whose p was < 0.05.

3. Results

The sample consisted of 12 shoulder rehabilitation sessions with 11 patients (n = 11) (5 men and 6 women) with different shoulder pathologies that limited their range of shoulder mobility. In one of the patients, both shoulders were treated. One patient performed two sessions for the rehabilitation of both shoulders, while the rest performed only one rehabilitation session. The ages ranged from 35 to 66 years (54.92 ± 3.16 years). Table 1 shows the data of the patients who participated in the testing of the device.

Table 1. Data of patients that tested the device. S11 and S12 were performed by the same patient, but with different shoulders (left and right, respectively).

Shoulder Session	Gender	Age	Pathology
1	M	66	Supraspinatus tendon rupture
2	F	45	Fracture with dislocation and calcified supraspinatus tendinosis
3	M	67	Complete rotator cuff rupture
4	M	57	Calcified supraspinatus tendinosis
5	M	46	Supraspinatus tendon suture due to subacromial syndrome
6	F	42	Supraspinatus tendinosis
7	M	66	Recovery after arthrolysis due to shoulder capsulitis
8	F	52	Supraspinatus tendon rupture
9	F	65	Rotator cuff tendinopathy
10	F	35	Supraspinatus tendinosis
11	F	61	Supraspinatus and infraspinatus rupture (left)
12	F	61	Tendinosis of the rotator cuff (right)

The analysis shows statistically significant changes for the ROM of external shoulder rotations. The mean and standard deviations of the external rotation of shoulder were 42.91 ± 4.53° for the pre-intervention measurements and 53.88 ± 4.26° for the post-intervention measurements. Figure 7 and Table 2 show the data and range of movement of the patients during the rehabilitation sessions.

The patient scores on the SUS questionnaire ranged from 65 to 97.5, with an average score of 82.70 ± 9.21, suggesting a high degree of acceptance. These scores are higher than 68, indicating that this device can be rated "excellent" in the acceptability range. Table 3 shows the means of the scores obtained for each question. Questions Q1, 3, 5, 7, and 9 were positive questions and the averages obtained in these questions were above 4. It is worth noting that Q1 "I think I would use this device frequently" presented an average score of 4.75 ± 0.59 and Q9 "I feel safe using this device" showed an average score of 4.91 ± 0.27.

Meanwhile, questions Q2, 4, 6, 8, and 10 were negative questions. The mean scores in these questions were 1, except in Q2 "I find this device unnecessarily complex" which had a mean score of 2.5 ± 1.55 and Q4 "I think I would need help from a person with technical knowledge to be able to use this device" with a mean score of 2.91 ± 1.49.

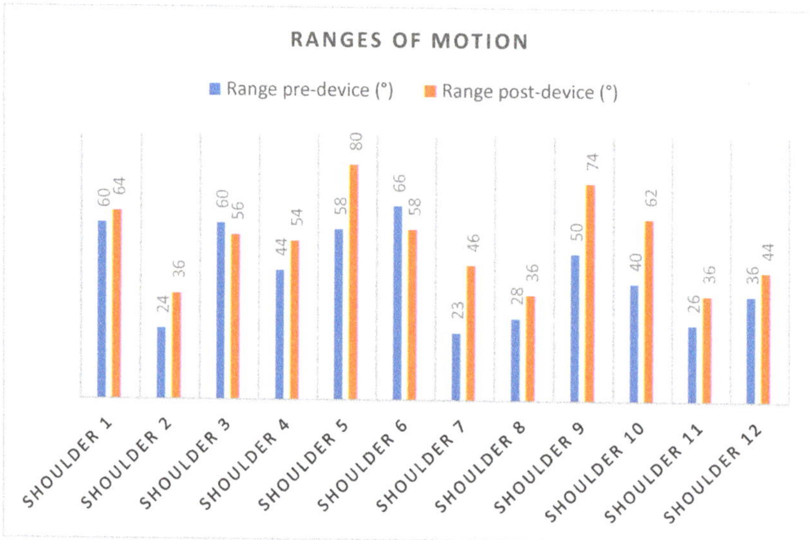

Figure 7. Ranges of motion after one rehabilitation session. A total of 10 out of 12 shoulders present improvements, while 2 present slight deteriorations, likely due to lack of familiarity with the device.

Table 2. Comparison of range of motion of shoulder external rotations.

Variable	Pre-Intervention Median (Interquartile Range)	Post-Intervention Median (Interquartile Range)	Intragroup Analysis Median *p*-Value
Range of motion Shoulder external rotations (degrees)	42.00 (33.00)	55.00 (25.50)	0.008 *

Data are expressed as the median and interquartile range. * $p < 0.05$ using the Wilcoxon test for related samples.

Table 3. Results of the system usability scale questionnaire.

Question	Score
Q1	4.75 ± 0.59
Q2	2.5 ± 1.55
Q3	4.66 ± 0.47
Q4	2.91 ± 1.49
Q5	4.08 ± 1.11
Q6	1.83 ± 1.28
Q7	4.58 ± 1.11
Q8	1.58 ± 1.32
Q9	4.91 ± 0.27
Q10	1.08 ± 0.27
TOTAL	80.71 ± 9.79

Data are expressed as mean ± standard deviation.

4. Discussion

The present study showed the use of an active device for shoulder rehabilitation based on a force control architecture. This preliminary study is based on a single treatment session to treat the external rotation movements of the shoulder in the patients in whom this was limited by musculoskeletal injuries. It also described the patient's acceptance of the device, and the results show a high level of approval from the patients after the first trial.

It has been demonstrated that robotic and haptic technologies, such as force control and real-time signal processing, have an effective contribution in upper limb rehabilitation. There are many examples of complex exoskeletons [15,16] and robot-like equipment applied in upper limb rehabilitation [17]. Physical therapy can be a long, lengthy, and costly process, which can lead to loss of interest in individuals undergoing therapy. Reducing recovery time is a potential solution that may motivate individuals to participate and continue physical therapy. Rehabilitation with an exoskeleton can reach more patients while reducing physician interactions and overall rehabilitation costs [18].

This equipment can perform a wide variety of movements but is specifically programmed for executing rehabilitation tasks. It is also necessary to design the proper patient interface to fix or attach the patient's upper limb to the robot structure. These kinds of solutions have the advantage of being able to adapt to different rehabilitation activities. Cost-effective automation devices, such as exoskeletons, are critical to the success of these interventions and rehabilitation programs.

With a single treatment session, this active device for shoulder rehabilitation based on force control was found to significantly increase the external rotation movement of the shoulder in the patients who used it. The pre-intervention mean and standard deviations of the ROM of the external rotation of the shoulder were $42.91 \pm 4.53^\circ$, and $53.88 \pm 4.26^\circ$ in the post-intervention measurements. In addition, patients reported high levels of acceptance of the device; their scores on the SUS questionnaire ranged from 65 to 97.5, with an average score of 82.70 ± 9.21, indicating a high degree of acceptance. These very promising results were obtained with minimal hardware and a user-friendly interaction with the patient.

This device is in line with a new generation of compact and simple rehabilitation equipment that is portable and easy to use for the patient [19], in contrast to the classical robotic solutions mentioned above. The main advantage of this kind of solution is the simplicity and cost-effective aspects. Only one degree of freedom is enough to perform shoulder rehabilitation, and this equipment can be adapted for horizontal and vertical rotations. Moreover, the reduced size of the device allows it to be easily transported to the rehabilitation workspace, which can be located in a clinic or at the patient's home.

This prototype is currently being improved in order to provide more functionalities to clinicians for performing new rehabilitation exercises and software is also being developed to better characterize the rehabilitation sessions.

5. Study Limitations

This study has several limitations. First, the small sample size cannot allow extrapolation of the results. Second, the ROM was evaluated with a manual goniometer. Futures studies are necessary using objective measures such as using a electrogoniometer or 3D motion analysis. Third, the sample was heterogeneous in respect to the injury and age. It is necessary to take this into account. Finally, the lack of follow-up does not allow us to conclude if the ROM improvement lasted over time.

6. Conclusions

In this paper, a compact and cost-effective prototype for shoulder rehabilitation was described. The most important features are the simplicity of the hardware and its effectiveness in rehabilitation exercises. The equipment is designed to repeat a shoulder rotation while reflecting a force on the patient's hand. This exercise allows muscle activation, while also increasing the range of shoulder movement.

This device allows patients to perform exercises for rehabilitation sessions that are an ideal complement to clinician manipulation. The preliminary results suggest that the use of this device and the incorporation of such equipment into rehabilitation services could be of great help for patients in their rehabilitation process and for physiotherapists in applying their therapies.

Author Contributions: Conceptualization, all authors; methodology, all authors; rehabilitation device design and development, David Pont-Esteban, J.C., P.E., M.F. and M.A.S.-U.; clinical validation, A.C.-G. and I.M.A.-D.; writing—review and editing, all authors. All authors have read and agreed to the published version of the manuscript.

Funding: This research did not receive external funding.

Institutional Review Board Statement: The study was carried out in accordance with the Declaration of Helsinki and approved by the Ethics Committee of the Universidad Rey Juan Carlos (N-20130081).

Informed Consent Statement: Informed consent was obtained from all subjects involved in the study.

Acknowledgments: The authors thank all the patients and clinicians who participated in the validation test of the exoskeleton.

Conflicts of Interest: The authors declare that they have no conflict of interest.

References

1. Lädermann, A.; Denard, P.J.; Collin, P. Massive rotator cuff tears: Definition and treatment. *Int. Orthop. (SICOT)* **2015**, *39*, 2403–2414. [CrossRef] [PubMed]
2. Abdulla, S.Y.; Southerst, D.; Cote, P.; Shearer, H.M.; Sutton, D.; Randhawa, K.; Varatharajan, S.; Wong, J.J.; Yu, H.; Marchand, A.A.; et al. Is exercise effective for the management of subacromial impingement syndrome and other soft tissue injuries of the shoulder? A systematic review by the Ontario Protocol for Traffic Injury Management (OPTIMa) Collaboration. *Man. Ther.* **2015**, *20*, 646–656. [CrossRef] [PubMed]
3. Lowry, V.; Lavigne, P.; Zidarov, D.; Perreault, K.; Roy, J.S.; Desmeules, F. Knowledge and appropriateness of care of family physicians and physiotherapists in the management of shoulder pain: A survey study in the province of Quebec, Canada. *BMC Prim. Care* **2023**, *24*, 49. [CrossRef] [PubMed]
4. Deslauriers, S.; Déry, J.; Proulx, K.; Laliberté, M.; Desmeules, F.; Feldman, D.E.; Perreault, K. Effects of waiting for outpatient physiotherapy services in persons with musculoskeletal disorders: A systematic review. *Disabil. Rehabil.* **2021**, *43*, 611–620. [CrossRef] [PubMed]
5. Ahmed, T.; Islam Md, R.; Brahmi, B.; Habibur Rahman, M. Robustness and Tracking Performance Evaluation of PID Motion Control of 7 DoF Anthropomorphic Exoskeleton Robot Assisted Upper Limb Rehabilitation. *Sensors* **2022**, *22*, 3747. [CrossRef] [PubMed]
6. Suárez Sanabria, N.; Osorio Patiño, A.M. Biomecánica del hombro y bases fisiológicas de los ejercicios de Codman. *Rev. CES Med.* **2013**, *27*, 205–217.
7. Teasell, R.; McIntyre, A.; Viana, R.; Bateman, E.A.; Murie-Fernandez, M.; Janzen, S.; Saikaley, M. Developing a framework for utilizing adjunct rehabilitation therapies in motor recovery of upper extremity post stroke. *Top. Stroke Rehabil.* **2023**, *30*, 493–500. [CrossRef] [PubMed]
8. Wu, J.; Cheng, H.; Zhang, J.; Yang, S.; Cai, S. Robot-Assisted Therapy for Upper Extremity Motor Impairment After Stroke: A Systematic Review and Meta-Analysis. *Phys. Ther.* **2021**, *101*, pzab010. [CrossRef] [PubMed]
9. Proietti, T.; Ambrosini, E.; Pedrocchi, A.; Micera, S. Wearable Robotics for Impaired Upper-Limb Assistance and Rehabilitation: State of the Art and Future Perspectives. *IEEE Access* **2022**, *10*, 106117–106134. [CrossRef]
10. Podobnik, J.; Mihelj, M.; Munih, M. Upper limb and grasp rehabilitation and evaluation of stroke patients using HenRiE device. In Proceedings of the 2009 Virtual Rehabilitation International Conference, Haifa, Israel, 29 June–2 July 2009; pp. 173–178. [CrossRef]
11. Zhao, Y.; Wu, H.; Zhang, M.; Mao, J.; Todoh, M. Design methodology of portable upper limb exoskeletons for people with strokes. *Front. Neurosci.* **2023**, *17*, 1128332. [CrossRef] [PubMed]
12. Norkin, C.; White, D.J. Manual de Goniometría. In *Evaluación de la Movilidad Articular*, 5th ed.; Editorial Paidotribo: Barcelona, Italy, 2019.
13. Brooke, J. SUS-A quick and dirty usability scale. *Usability Eval. Ind.* **1996**, *189*, 4–7.
14. Brooke, J. SUS: A retrospective. *J. Usability Stud.* **2013**, *8*, 29–40.
15. Kadivar, Z.; Beck, C.E.; Rovekamp, R.N.; O'Malley, M.K. Single limb cable driven wearable robotic device for upper extremity movement support after traumatic brain injury. *J. Rehabil. Assist. Technol Eng.* **2021**, *8*, 20556683211002448. [CrossRef] [PubMed]
16. Gupta, S.; Agrawal, A.; Singla, E. Architectural design and development of an upper-limb rehabilitation device: A modular synthesis approach. *Disabil. Rehabil. Assist. Technol.* **2022**. [CrossRef] [PubMed]

17. Hailey, R.O.; De Oliveira, A.C.; Ghonasgi, K.; Whitford, B.; Lee, R.K.; Rose, C.G.; Deshpande, A.D. Impact of Gravity Compensation on Upper Extremity Movements in Harmony Exoskeleton. In Proceedings of the 2022 International Conference on Rehabilitation Robotics (ICORR), Rotterdam, The Netherlands, 25–29 July 2022; pp. 1–6. [CrossRef]
18. Delgado, P.; Yihun, Y. Integration of Task-Based Exoskeleton with an Assist-as-Needed Algorithm for Patient-Centered Elbow Rehabilitation. *Sensors* **2023**, *23*, 2460. [CrossRef] [PubMed]
19. Garzo, A.; Jung, J.H.; Arcas-Ruiz-Ruano, J.; Perry, J.C.; Keller, T. ArmAssist: A Telerehabilitation Solution for Upper-Limb Rehabilitation at Home. *IEEE Robot. Autom. Mag.* **2003**, *30*, 62–71. [CrossRef]

sensors

MDPI

Article

An End-to-End Dynamic Posture Perception Method for Soft Actuators Based on Distributed Thin Flexible Porous Piezoresistive Sensors

Jing Shu [1], Junming Wang [1], Kenneth Chik-Chi Cheng [2,3], Ling-Fung Yeung [1], Zheng Li [1,4,*] and Raymond Kai-yu Tong [1,*]

1 Department of Biomedical Engineering, The Chinese University of Hong Kong, Hong Kong SAR 999077, China; 1155138492@link.cuhk.edu.hk (J.S.); 1155151867@link.cuhk.edu.hk (J.W.); lfyeung@cuhk.edu.hk (L.-F.Y.)

2 Department of Biomedical Engineering, The Hong Kong Polytechnic University, Hong Kong SAR 999077, China; kenneth-c.cheng@polyu.edu.hk

3 Research Institute for Sports Science and Technology, The Hong Kong Polytechnic University, Hong Kong SAR 999077, China

4 Department of Surgery, The Chinese University of Hong Kong, Hong Kong SAR 999077, China

* Correspondence: lizheng@cuhk.edu.hk (Z.L.); kytong@cuhk.edu.hk (R.K.-y.T.)

Abstract: This paper proposes a method for accurate 3D posture sensing of the soft actuators, which could be applied to the closed-loop control of soft robots. To achieve this, the method employs an array of miniaturized sponge resistive materials along the soft actuator, which uses long short-term memory (LSTM) neural networks to solve the end-to-end 3D posture for the soft actuators. The method takes into account the hysteresis of the soft robot and non-linear sensing signals from the flexible bending sensors. The proposed approach uses a flexible bending sensor made from a thin layer of conductive sponge material designed for posture sensing. The LSTM network is used to model the posture of the soft actuator. The effectiveness of the method has been demonstrated on a finger-size 3 degree of freedom (DOF) pneumatic bellow-shaped actuator, with nine flexible sponge resistive sensors placed on the soft actuator's outer surface. The sensor-characterizing results show that the maximum bending torque of the sensor installed on the actuator is 4.7 Nm, which has an insignificant impact on the actuator motion based on the working space test of the actuator. Moreover, the sensors exhibit a relatively low error rate in predicting the actuator tip position, with error percentages of 0.37%, 2.38%, and 1.58% along the x-, y-, and z-axes, respectively. This work is expected to contribute to the advancement of soft robot dynamic posture perception by using thin sponge sensors and LSTM or other machine learning methods for control.

Keywords: soft robotics; soft sensors; flexible porous structures; deep learning; long short-term memory

Citation: Shu, J. ; Wang, J.; Cheng, K.C.-C.; Yeung, L.-F.; Li, Z.; Tong, R.K.-y. An End-to-End Dynamic Posture Perception Method for Soft Actuators Based on Distributed Thin Flexible Porous Piezoresistive Sensors. *Sensors* **2023**, *23*, 6189. https://doi.org/10.3390/s23136189

Academic Editors: Xin Zhao, Mingzhu Sun and Qili Zhao

Received: 23 May 2023
Revised: 30 June 2023
Accepted: 3 July 2023
Published: 6 July 2023

1. Introduction

Soft robots, a novel category of robots that has emerged in recent decades, are characterized by their high flexibility, compliance, and safety [1]. Compared with conventional rigid robotics, soft actuators and robotic devices have advantages in human–robot interaction, range of motions, fabrication difficulties, and adaptations to varying environments [2]. However, the perception of posture, which can be easily represented by the status of joints and links in rigid-link robots, presents a significant challenge in the development of soft robotics due to the difficulty in accurately modeling elastic continuous structures.

Researchers have developed sensing methods, which can be broadly categorized into two main groups: on-board sensing and external sensing of robot posture. On-board sensors, such as flexible bending gauges [3,4] and inertial measurement units (IMUs) [5,6], may lack flexibility and restrict the robot's motion. Fully soft sensors embedded in flexible materials [7–10] require complex fabrication methods or specialized equipment. To overcome these challenges, researchers have also utilized external sensing methods, such as vision-based sensing [11–13] and magnetic tracking [14–16], to obtain posture information for soft robots, particularly for proof-of-concept use in laboratory environments.

As a result, there is an increasing need to develop advanced and flexible sensing techniques that can accurately detect the postures of soft robots. In [17,18], researchers developed a piezoresistive actuator body by mixing carbon fiber with elastomer, resulting in changes in resistance values during deformation. In [19], researchers achieved soft robot posture sensing by utilizing a kirigami-inspired sensor and a deep learning calibration method, representing an innovative approach to the closed-loop control of soft robots. In the recent work of Ye et al. [20], a significant breakthrough has been achieved in robotics by utilizing a kirigami-inspired porous sensor for robot teleoperation. In addition to their use in detecting robot postures, flexible sensing techniques have found extensive applications in other areas that require safe and comfortable human–robot interactions, such as the development of artificial skins [21,22] and the monitoring of human body conditions and motions [23–25]. In this paper, we propose an end-to-end method that effectively captures the dynamic behavior of soft actuators, with potential applications for perceiving the posture and dynamic behavior of objects during deformation, such as bellow-shaped soft actuators, where deformation is mainly driven by buckling [26]. Specifically, the neighbor wall angle α (see Figure 1) changes during actuator deformation, allowing us to indirectly perceive the posture of the soft robot through the observation of angle changes.

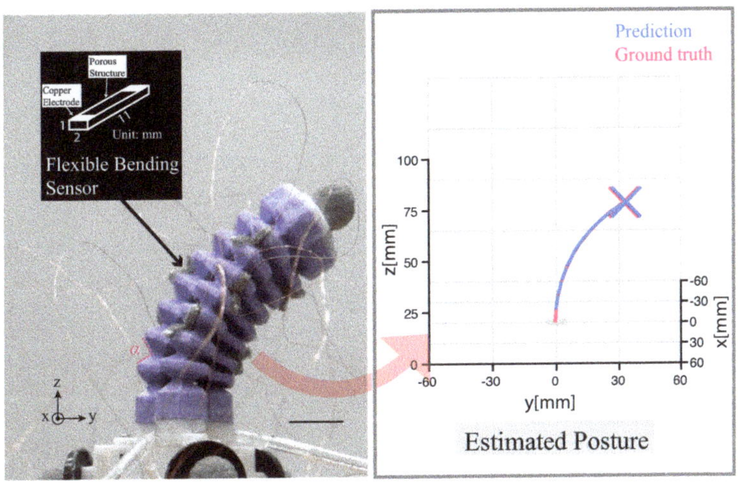

Figure 1. Proprioception of posture in a pneumatic soft actuator. On the side of the actuator (left), there are nine distributed flexible bending sensors (scale bar = 1 cm) made by a thin conductive sponge via the planar fabrication method. The neighbor wall angle α was indicated. A trained neural network was employed to calibrate the sensor, and the 3D representation of the posture of soft actuator is shown on the right. (The perspective of the right figure is a top–down view along the x−z plane, parallel to the vector (3, 0, 1)).

In this study, we utilized commercially available thin conductive sponges as the sensor material, which were attached to the lateral surface of a three-degree-of-freedom (DoF) soft actuator. Previous research has demonstrated that the piezoresistive properties of thin conductive sponges enable the detection of deformation and tactile information of attached objects [22,27–33]. Due to their flexible and lightweight nature, sensors developed with sponge materials have been widely employed in human–robot interactions, including tactile sensing [22,27,29,30] and human motion detection [28]. The porous and flexible nature of the conductive sponge structure minimizes impediments to the movement of attached objects induced by sensor deformation.

Although recent research work has explored the feasibility of collecting the tactile information of robots using conductive sponge sensors [31,34,35], only a few studies have employed sensors developed with conductive sponge material to detect or estimate the deformation of soft robots. In our previous work [36], we proposed a kirigami-inspired flexible sponge sensor that could estimate the bending angle of a single degree-of-freedom flexible fiber-reinforced bending actuator. By connecting the sensor information and calibration neural network organically, we achieved highly accurate robot posture estimation. In this paper, we simplified the sensor structure to make it suitable for bending detection and increased the sensor number, enabling us to achieve dynamic 3D posture perception of soft robots. To the best of our knowledge, this is the first study to seamlessly combine sponge sensors and a neural network calibration method to achieve soft robot dynamic 3D posture perception.

We employed an end-to-end method in our study to calibrate our sensing system via RNN deep learning. This approach obviates the need for developing a complex kinematic model. The idea has been widely explored and validated in the soft robotics community [9,19,36–39] as a means of addressing the non-linearity and hysteresis issues associated with data. We utilized a long short-term memory (LSTM) neural network [40], which is a specialized type of recurrent neural network (RNN) that can handle sequential data with the aforementioned difficulties. We trained an LSTM neural network using feedback sensing signals from nine sensors distributed on the side wall of the actuator. The posture of a 3-DoF soft actuator was predicted using this network in conjunction with the input sensor value, and the overall root mean square error (RMSE) value was calculated to gauge the accuracy of our sensing system. Despite the acceptable prediction accuracy of the sensor, certain issues were uncovered during the study. The primary issue concerns the saturation of sensor readings when the sensor approaches full compression, thereby leading to a loss of prediction accuracy in situations where the attached object undergoes extreme deformation, which would be indicated in the experimental results of sensor characterization.

Overall, this paper makes the following contributions:

(1) An end-to-end (indirect) method for soft robot posture perception with flexible porous piezoresistive sensors via the RNN deep learning approach.
(2) The design and characterization of a scalable miniature flexible bending sensor that can be quickly attached to existing bellow-shaped actuators to perceive the posture and dynamic behaviors as sensing units while minimizing the effect of motion restriction.
(3) The validation of our proposed method on a small-scale 3 DoF bellow-shaped soft robot actuator with three pneumatic chambers.

2. Design Overview and Rationale

2.1. Soft Robot and Flexible Porous Piezoresistive Sensor Design

Fluidic elastomer actuators (FEAs) are widely used in soft robotics. To perceive the posture of FEAs, researchers have employed various methods, including flexible bending gauges [3], sensors with micro-channels and liquid metal [9], piezoresistive materials embedded in the actuator body [41], fiber optic intensity-based sensors [42], and others. As mentioned in Section 1, these sensing methods can either restrict the motion of soft actuators or require bulky serving systems. Consequently, researchers have pursued a sensing method that can integrate independent sensors into the FEA body. Such a method was first developed in [19], which bypassed the need for designing both the soft actuators and their sensing systems. In this study, we present a robot sensing system that utilizes a similar distributed sensing method, employing bending-based thin conductive sponge sensors. Our approach can be easily applied to the surface of soft actuators that undergo deformation during changes in posture.

We designed and fabricated a negative pressure-powered bellow-shaped FEA. The FEA body was fabricated using casting techniques and silicon material (Smooth-Sil 945,

Smooth-On Inc., East Texas, PA, USA). The actuator comprises three hexagonal bellow-shaped chambers, with hex-edge lengths of 5 mm, bonded in parallel using silicon adhesive (Sil-Poxy, Smooth-On Inc., USA). The angle between two bellow walls is 55 degrees, and each chamber was constructed by bonding two half-chambers and two sealing caps with the same silicone adhesive. The fabrication process of the FEA is illustrated in Figure 2c–e.

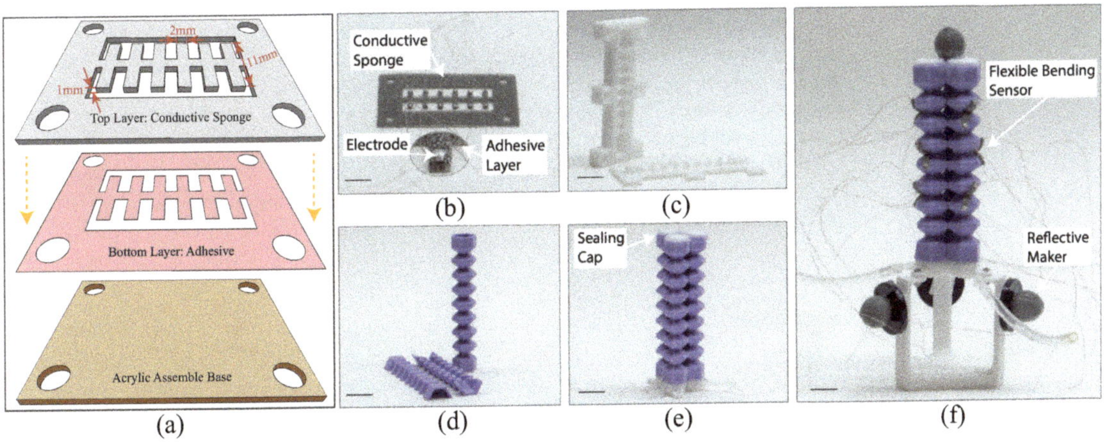

Figure 2. Fabrication of flexible bending sensor and pneumatic soft actuator. (**a**) Sketches of planer fabrication process of sensor body. (**b**) Flexible bending sensor via planar fabrication method. (**c**) Molds for silicone casting. (**d**) Half-chamber of the soft actuator made by silicon after the demolding process. Two half-chambers were combined to form one air chamber using silicon adhesive. (**e**) Three air chambers combined via silicon adhesive with top and bottom caps. (**f**) Soft pneumatic actuator with distributed flexible bending sensors. Sensors were attached to the actuator side via silicon adhesive. Four reflective makers are used to determine the true posture of the soft actuator (scale bar = 1 cm).

The planar fabrication method has been applied in the sensor fabrication, similar to the methods used in [36,43]. By employing such manufacturing techniques, it is possible to produce the sensors in small batches. The sketches of the planer fabrication process are shown in Figure 2a. Each flexible bending sensor comprises two layers fabricated using planar fabrication methods and cut by a laser cutter. The top layer is made of a 1 mm thin layer and 2 mm wide PU conductive sponge material (Beilong Inc., Wujiang, China) with a constant thickness, while the bottom layer is composed of adhesive (300LSE, 3M, Saint Paul, MI, USA). Copper electrodes with leading wires were attached to the two ends of the sponge using silver conductive adhesives (3703, Sinwei, Shanghai, China) (Figure 2b). After drying at room temperature for 12 h, the sensors were trimmed and the undesired portions were removed. The sensors were then affixed to the actuator body using the aforementioned silicon adhesive (Figure 2f). To fit the finger-sized bellow-shaped actuator, the length, width, and thickness of the flexible bending sensor were selected to be 11 mm, 2 mm, and 1 mm, respectively. The length and width of the flexible bending sensors can be varied by changing the laser cutting pattern to match the dimensions of different soft actuators, while the thickness of the sensors can be adjusted by selecting a conductive sponge sheet with a different thickness. The number of sensors attached to the soft actuator depends on the required level of accuracy. In our design, we bonded nine thin sponge sensors to the actuator body.

2.2. Working Mechanism of Flexible Porous Piezoresistive Sensor

The resistance of our thin sponge sensor follows the relationship $R_s = C(l/A)$ since porous structures are evenly distributed inside, where C is a constant relative to the resistivity and porosity of the material. l and A represent the length and cross-sectional area of the sensor, respectively. The schematic diagram of the flexible sponge sensor and its operative principle is depicted in Figure 3. The left-hand side of the figure illustrates the fundamental components of the flexible sponge sensor. The electrical resistance of the sensor is sensitive to the angle of the attached object, as denoted by the variable α as shown on the right-hand side of Figure 3. By subjecting the attached object to external deformations, the angle α is likely to vary, thereby causing a change in the compression region of the sensor. This alteration in the compression region affects the electrical resistance of the sensor. Hence, the resistance of the flexible sponge sensor provides an indirect measure of the posture or deformation of the attached object. It is noteworthy that the object under investigation in this study is the bellow-shaped FEA. Developing a mathematical model for the sponge sensor would be inefficient due to the uncertain nature of its porous structure. Therefore, a learning-based method is more appropriate, and we utilized an LSTM neural network as our calibration neural network. The design of our LSTM neural network is presented in Section 2.4.

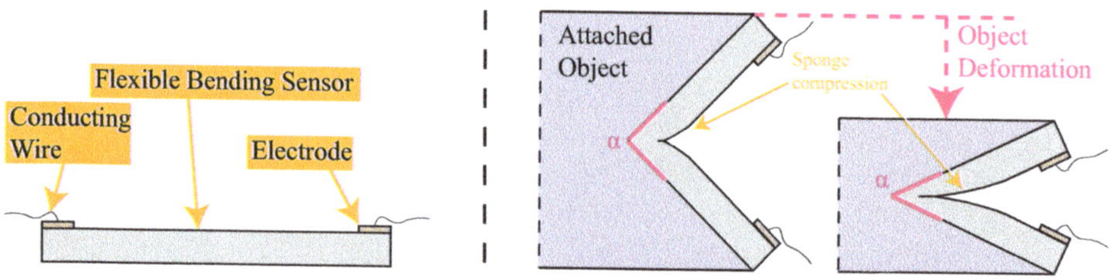

Figure 3. The cross-view sketch in the left panel shows the flexible sponge sensor, while its working mechanism is depicted in the right panel. When there is deformation on the attached object, the sponge compression region changes, resulting in a change in electrical resistance.

2.3. *Soft Robot Kinematic Description*

We utilized the parameter vector $\mathbf{q} = (k, \phi, L)$ to represent the posture of the soft actuators (Figure 4), where k is the curvature of this bellow-shaped FEA. Here, we used a constant-curvature model to simplify the problem. Therefore, we have $k = 1/r$, where r is the radius of the actuator arc. ϕ represents the angle between the $x - z$ plane of the reference frame and the plane of the actuator arc. L represents the length of the actuator, which is relative to the length of the three parallel chambers l_1, l_2, and l_3. For our actuator, the length of each chamber l_i $(i = 1, 2, 3)$ is relative to β, which is the angle between neighboring walls. For simplicity, we have $l_i = n \cdot a \cdot \sin(\beta/2)$, where n is the number of bellow units, and a is the length of the bellow wall segments.

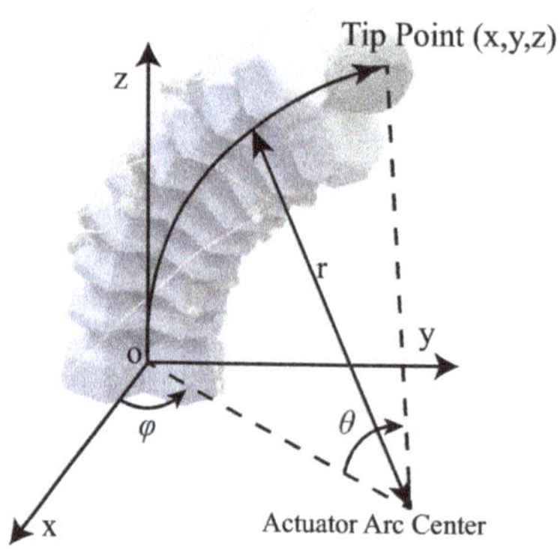

Figure 4. Kinematic representation of soft pneumatic actuator. The principle axis of reference frames was indicated.

In our case, we can directly measure the coordinates of the tip point $P_1 = (x_1, y_1, z_1)$ and the rear point P_2 in the task space of the actuator. To transfer these coordinates to the configuration space, some manipulations are required. We set P_2 as the origin, and then ϕ can be calculated based on the tangent values of x_1 and y_1 if the projection of the actuator on the $x - y$ plane does not lie on the primary axis.

$$\phi = \arctan\left(\frac{y_1}{x_1}\right) \tag{1}$$

When the tip is located on the z-axis, the actuator is at singularity. In this case, we set ϕ to be zero.

Then, we rotate the actuator arc to the $x - z$ plane by multiplying a rotational matrix about the z-axis. Then, we could obtain $P_1' = (x_1, y_1, z_1)$ on the $x - z$ plane,

$$\begin{bmatrix} x_1' \\ y_1' \\ z_1' \end{bmatrix} = \begin{bmatrix} \cos(-\phi) & -\sin(-\phi) & 0 \\ \sin(-\phi) & \cos(-\phi) & 0 \\ 0 & 0 & 1 \end{bmatrix} \begin{bmatrix} x_1 \\ y_1 \\ z_1 \end{bmatrix} \tag{2}$$

After that, we have

$$\begin{aligned} x_1' &= r(1 - \cos\theta) \\ z_1' &= r\sin\theta \end{aligned} \tag{3}$$

Solving Equation (3), we can obtain the values of r and θ. Then, the actuator length L can be calculated using $L = r \cdot \theta$. Reference [44] provides more details on posture representations. In the rest of the paper, we will use either the parameter vector $\mathbf{q}$ or the tip point coordinates P_1 to represent the posture of the bellow-shaped FEA. We will use an LSTM neural network as a function approximator for the time-series response of sensors. The velocity, i.e., dynamic behavior, of the tip point of the actuator can be determined by taking the first-order differentiation of the tip coordinates, which is suitable for both the predicted value and the ground truth.

2.4. Neural Network Design

We employed an LSTM neural network to establish the relationship between the values of the thin sponge sensor and the posture of the bellow-shaped actuator. As previously discussed in Section 2.1, the development of an accurate mathematical model for the sponge sensor is an inefficient process. Hence, a learning-based approach was deemed optimal for our purposes.

The neural network architecture comprises an input layer, a fully connected layer, and an output layer, with the number of hidden layers ranging from one to four (validation results with different numbers of hidden layers are compared in Figure 5). Dropout layers follow the hidden layers to prevent overfitting. The inputs consist of a time-series sensing signal feedback from each sensor (R1, R2, R3... R9), and the output represents the absolute spatial position (x, y, z). The LSTM neural network's structure is illustrated in Figure 6. The use of dropout layers helps to prevent overfitting. Furthermore, fine-tuning hyperparameters such as the number of hidden layers, number of hidden states, and dropout layer rate is critical to achieving optimal performance, which is compared and determined in Section 4.2. The overall root mean square error (RMSE) is selected as the cost function to evaluate the performance of the trained LSTM neural network, representing its ability to perceive the 3D dynamic posture of the bellow-shaped FEA in the presence of non-linear and hysteretic sensor feedback response. The MATLAB Deep Learning Toolbox is utilized to train and test the LSTM neural network, as detailed in Section 3.2.

Hidden neurons \ RMSE \ Dropout rate	0.10	0.25	0.50
10	20.62	23.68	18.85
25	13.39	13.15	12.93
50	14.20	13.64	12.75
100	13.96	14.01	12.57

(a) One Hidden Layer

Hidden neurons \ RMSE \ Dropout rate	0.10	0.25	0.50
10	16.89	17.68	19.33
25	16.00	13.41	12.91
50	16.28	13.49	12.90
100	15.01	14.51	14.23

(b) Two Hidden Layers

Hidden neurons \ RMSE \ Dropout rate	0.10	0.25	0.50
10	14.99	17.19	15.94
25	15.18	15.12	12.81
50	17.18	16.51	17.49
100	14.59	15.93	14.35

(c) Three Hidden Layers

Hidden neurons \ RMSE \ Dropout rate	0.10	0.25	0.50
10	16.31	17.35	16.40
25	15.77	13.11	14.48
50	15.30	15.62	13.70
100	13.85	19.81	14.04

(d) Four Hidden Layers

Figure 5. RMSE of validation set prediction for LSTM neural network with different neural network parameters (unit: mm).

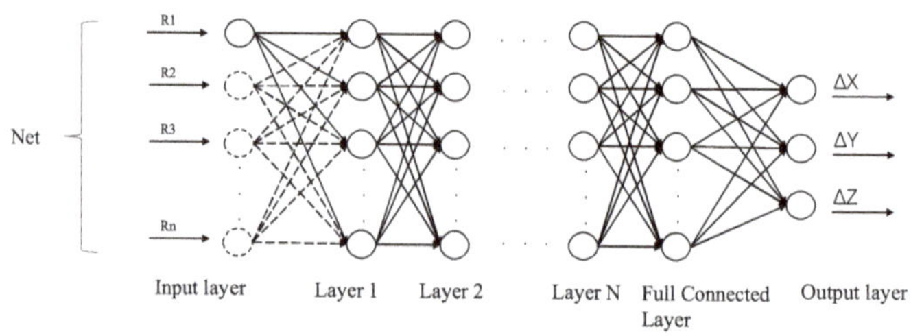

Figure 6. Structure of LSTM neural network. The neural network contains the input layer, hidden layer, fully connected layer, and output layer.

3. Experiment Setup

3.1. Sensor Response Experiment Setup

In this experiment, we investigated the electrical resistance and rotating torque response of a thin sponge bending sensor as a function of the bending angle. One end of the sensor was affixed to a fixed plate, while the other end was attached to a rotational plate with a virtually fixed axis of rotation. The experimental setups for the electrical resistance and sensor torque response tests are illustrated in Figure A1a,b, respectively. In the experiment, to measure the electrical resistance response of the sensor, the rotational plate was driving by a worm gear and stepper motor, which allow the plate to rotate with a speed of 4°/s accurately. To mitigate the impact of fabrication defects, an enlarged sensor with dimensions of 30 mm in length, 5 mm in width, and 2 mm in thickness was employed for the resistance measurement test. The sensor was connected in series with a 220 Ω resistor in the measuring circuit. The value of angle β varied between 10° and 100° periodically for the diverse test. The electrical resistance of the sensor during the first ten cycles was recorded and calculated during the diverse cycle test. The average electrical resistance of the sensor when values of β are round numbers of ten are calculated with mean and standard deviation (S.D.) value. For the sensor torque testing setup, a torque sensor with a measuring range of ±0.1 Nm (SBT850A, Simbatouch, China) was positioned at the bottom of the rotational plate. Both the enlarged sensor and the sensor attached on the soft actuator were tested. The electrical conducting wires were removed to eliminate the effects of cable tension on the testing results. The torque was recorded once it reached a stable state. The sensor torque was measured in increments of 10°.

3.2. Soft Robot Operation and Data Acquisition

The experimental setup for soft robot operation and data acquisition included a vacuum pump that generated negative pressure, an MCU (Portenta H7, Arduino, Ivrea, Italy) equipped with three proportional valves (ITV2090-212L, SMC Corporation, Tokyo, Japan) to regulate the pressure inside each actuator chamber, a data acquisition card (USB-6212, National Instrument, Austin, USA), and a motion capture system (Bonita 10, Vicon Inc., Yarnton, UK).

To characterize the robot's performance, the pressure inside each chamber was sequentially varied from 0 to −70 kPa at a rate of 2.44 kPa/s to explore the soft actuator's working space. A comparison was made between soft actuators with and without sensors. During the data acquisition period for neural network training, the pressure within each actuator chamber was randomly varied within the range of 0 to −70 kPa with a 25 Hz refresh rate. Within each refreshing period, the pressure changed randomly from 0 to −70 kPa. Consequently, the pressure within each actuator exhibited a random triangular wave pattern.

In the data acquisition circuit, the resistance R_s of the thin sponge sensors was converted into voltage information V_{out} using voltage divider circuits with the relationship $V_{out} = V_{cc}R_s/(R_0 + R_s)$, where V_{cc} and R_0 are 0.5 V and 20 Ω, respectively. The sensor information V_{out} was collected by a DAQ card at a frequency of 1000 Hz, which was sufficient for capturing the motion of pneumatic soft robots. During data processing, the sensor data was downsampled to a frequency of 100 Hz. A low-pass filter with a cut-off frequency of 20 Hz was applied to the sensor data to remove high-frequency noise.

The Vicon motion capturing system employed nine cameras distributed on the ceiling to capture the coordinates of reflective markers at a frequency of 100 Hz. The diameter and mass of the sphere markers were 1 cm and 0.42 g, respectively, and they did not significantly affect the performance of the soft actuator. One marker was placed on the top of the actuator to obtain the coordinates of the top point (P_1), while three markers were placed on the fixed rear part to obtain the coordinates of the rear point (P_2). The overall measuring error of the motion capturing system was less than 0.5 mm.

3.3. Neural Network Training

The data used to train the LSTM network were collected from random movements, with the pressure in the three chambers being randomly varied between 0 and −70 kPa. The training and validation data were obtained within 10 min and 400 s, respectively, with 70 s of validation data presented in Section 4.

The LSTM network training and generation were performed using the MATLAB Deep Learning Toolbox, with the default RMSE cost function and the Adam optimizer algorithm [45]. L2 regularization was adopted to prevent overfitting of the data. The initial learning rate was set to 0.001, and network performance was validated using a test set every 30 iterations, with overall iteration stopping after 100 epochs. The time-series data were sequentially divided, with 85% of the samples assigned to the training set and 15% assigned to the test set. The time series spanned 150 s and contained 9000 data points. Training the network with different sets of hyperparameters, including the number of hidden layers, number of hidden states, and dropout layer rate, was critical to optimizing network performance. These parameters were varied in the ranges [1, 2, 3, 4], [25, 50, 100, 200], and [0.1, 0.25, 0.5]. The root mean square error (RMSE) was used to evaluate the prediction accuracy, with $\texttt{RMSE} = \sqrt{\texttt{mean}[(\texttt{error}^*)^2]}$, where $\texttt{error}^* = \sqrt{(x_{pred} - x_{true})^2 + (y_{pred} - y_{true})^2 + (z_{pred} - z_{true})^2}$ represents the distance between the predicted and true actuator tip positions. The results of these calculations are summarized in Section 4.2. For the neural network with the lowest RMSE value, the neural network parameters were employed to estimate the data in the validation dataset, which is discussed in Section 4.3.

4. Experiment Results

4.1. Soft Sensor and Actuator Characterization

4.1.1. Soft Actuator Working Space

In this experiment, the response characteristics of the soft actuator and sensors were explored, and the working space of the actuator with and without thin sponge bending sensors was determined. The tip point was treated as the origin in the experiment to facilitate comparison. When the pressure inside the three air chambers was varied from 0 to −70 kPa, the bending angle θ (as shown in Figure 4) varied between 0 and 90.69 degrees, while the actuator length L varied between 80.2 and 64.2 mm. Figure 7 shows the coordinates of the actuator tip point during changes in air pressure inside the chambers. The red curve represents the trajectory of the tip point when sponge bending sensors are attached to the actuator, while the blue curve represents the tip point trajectory when no sponge bending sensor is attached. These two working spaces are almost identical. Therefore, we can conclude that the attachment of distributed sponge bending sensors does not significantly affect the working space of the soft actuator.

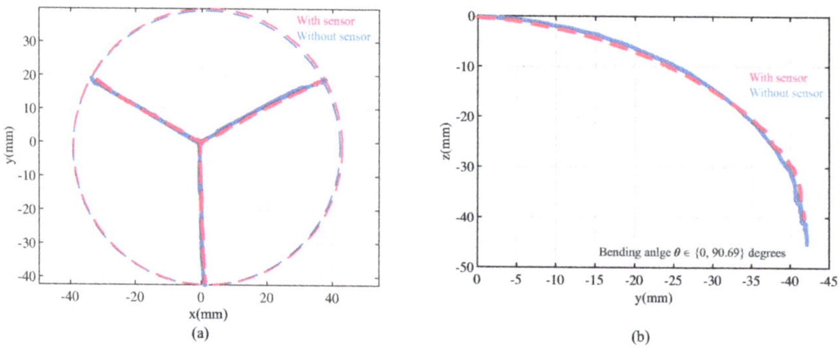

Figure 7. Working space of bellow−shaped actuator with (red curve) and without (blue curve) sponge bending sensor. (**a**) Top view of working space. (**b**) Side view of working space. The bending angle θ of the actuator could vary between 0 and 90.69 degrees. This shows that the thin sponge sensor has negligible influence on the soft actuator's motion.

4.1.2. Response of Flexible Bending Sensor

The electrical resistance response of the sensor is presented in Figure 8a. The small cycles on the figure represent the mean value of the electrical resistance during the ten cycles test when the value of angle β is the round numbers of ten. The error bars are representing the correspondence standard deviation (S.D.) value. The resistance of the enlarged sponge bending sensor varied from 8 to 31 Ω as the angle between two rigid plates β changed from 10 to 100 degrees. The sensor's resistance in the ten-cycle diverse test would go up gradually (Figure 9). Calibrating such sensor drifting using the conventional model-based calibration method would be challenging. In this work, by combining the flexible sensors and LSTM neural network calibration method, we can overcome the sensor drifting issue and predict the posture of the soft actuator with acceptable accuracy, which is presented Section 4.2. The sensitive region of the sponge bending sensor with these dimensions was found to be between 10 and 60 degrees. When the angle β exceeded 60 degrees, the sensor became less sensitive, while when the angle β was less than 10 degrees, most parts of the sensor were fully compressed, leading to saturation, as also shown in Figure 10. The sensitive region of the sponge bending sensor may vary depending on the sensor dimensions, but it can typically be divided into the aforementioned three sections.

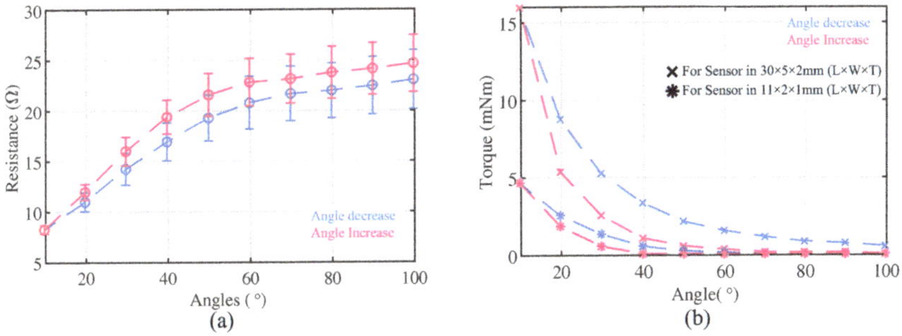

Figure 8. Experimental results of flexible bending sensor in (**a**) electrical resistance response test and (**b**) sensor torque response test. The responses of sensors when the bending angle decreased and increased are indicated by the blue line and red line, respectively.

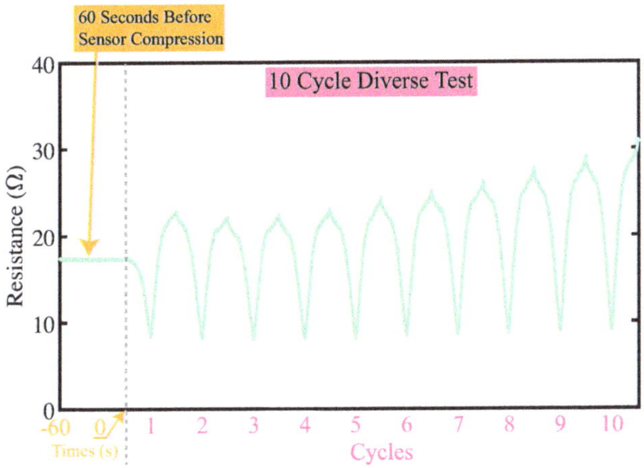

Figure 9. Diverse test results. Resistance of sensor during the 10−cycle diverse test together with 60 s data before test are presented to indicate sensor responses.

As discussed in Section 3.2, the response of the flexible bending sensor when attached to soft pneumatic actuators was recorded, and the resulting voltage changes were plotted in Figure 10. Sensors #1–3, #4–6, and #7–9 were attached to the same bellow side wall from bottom to top, as shown in Figure 10a, resulting in similar responses of the three sensors with respect to actuator shape-changing. In this experiment, we determined the saturation region of the thin sponge bending sensor. When the air pressure was lower than −50 kPa, the voltage change due to actuator deformation was small and smooth (Figure 10b). Beyond this range, the sensor output saturated as the sponge edges fully contacted each other. This phenomenon is further discussed in Section 4.3.

Figure 8b depicts the sensor torque response of the sensor as a function of the sensor angle β. The torque exhibits an inverse relationship with respect to the angle β, with the blue and red lines representing the response of the sensor during angular decreases and increases, respectively. The maximum torque values for the enlarged sensor and the sensor attached to the actuator were found to be 16 mNm and 4.7 mNm, respectively. These results support the notion that the attached sensor does not significantly affect the working range of the bellow-shaped actuator, as discussed in Section 4.1.1.

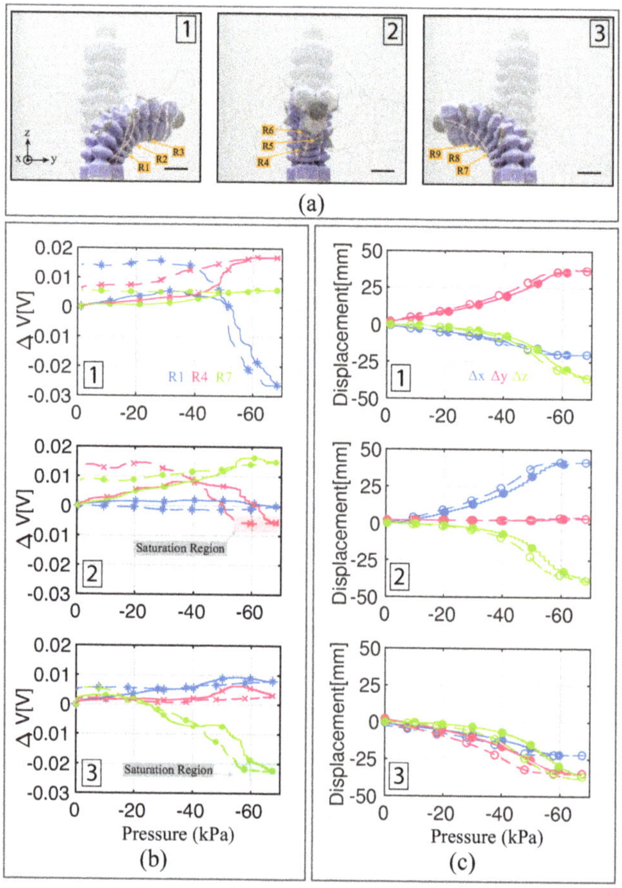

Figure 10. Characterization of thin sponge bending sensor integrated pneumatic soft actuator. (**a**) Photos show that −70 kPa negative pressure was applied to air chambers 1, 2, and 3, respectively (scale bars = 1 cm). (**b**) Sponge bending sensor (#1, #4 and #7) response (voltage changing) with respect to pressure changing in three air chambers with an interval of 1 kPa. (**c**) Tip displacement of actuator versus deflation pressure applied to each air chamber with same actuation pattern shown in (**a**). The tip point was treated as the original point. Index 1, 2, and 3 in each figure indicating each actuated air chamber. In both (**b**) and (**c**), filled lines and dashed lines are representing bending and relaxing processes, respectively. A median filter was applied in the plotting of (**b**).

4.2. Calibration Neural Network Performance

As delineated in Section 3.3, the process of optimizing the LSTM calibration neural network entailed fine-tuning three key parameters: the quantity of hidden layers, the number of hidden neurons, and the dropout rate in dropout layers. To gauge the performance of these neural networks, root mean square error (RMSE) values corresponding to the predicted tip point coordinates were computed, and the results are presented in Figure 5. An analysis of Figure 5 reveals that the LSTM neural network configuration with a single hidden layer, 100 hidden neurons, and a dropout rate of 0.5 yields the most favorable RMSE value of 12.57 mm. Subsequently, this LSTM neural network will be utilized to validate the 3D posture prediction of the soft actuator.

4.3. Validation of 3D Posture and Dynamic Behavior Prediction

Using the trained LSTM neural network discussed in Section 4.2, we were able to predict the 3D postures and dynamic behaviors of soft actuators, as demonstrated in Figure 11. The validation results presented in Figure 11a show three moments captured with time steps in the actuator random motions. Figure 11b presents 3D representations of actuator postures plotted with prediction and ground truth. For each moment, we calculated and listed parameter $\mathbf{q}$ based on the algorithm described in Section 2.3. Throughout the validation period, the predicted and ground truth displacements of the tip points are shown in Figure 11c, with displacements in the x-, y-, and z-axis directions represented by blue, red, and green colors, respectively. The dashed lines represent the ground truth of displacement in each direction. The RMSE values of the predicted value in the x-, y-, and z-axis directions were found to be 7.37 mm, 6.35 mm, and 3.70 mm, respectively. To assess the accuracy of the prediction, we evaluated the mean value and standard deviation (SD) value of prediction error in the coordinate axis direction, as reported in Table 1. The error percentage was computed by the absolute values of mean prediction error value divided by the working space of the actuator in the x-,y-, and z- axis direction. It is worth noting that the primary source of error was the saturation region, which can occur when the bending angle of the soft actuator is too large, causing the sensor to enter the saturation region. In our study, we found that a minimum of three sensors was required to estimate the robot's posture. We present the posture estimation results for sensors arranged at low (sensors #1, #4, and #7), middle (sensors #2, #5, and #8), and high (sensors #3, #6, and #9) levels (shown in Figure 10) in Table 2.

Table 1. Prediction accuracy of actuator tip position for validation data set.

Direction	Prediction Accuracy (mm)	Error Percentage
x-axis	-0.31 ± 7.37	0.37%
y-axis	2.00 ± 6.03	2.38%
z-axis	1.33 ± 3.46	1.58%

Table 2. Prediction accuracy of actuator tip position for validation data set using single-layer thin sponge flexible sensors.

Sensors	Prediction Accuracy in x-Axis (mm)	Prediction Accuracy in y-Axis (mm)	Prediction Accuracy in z-Axis (mm)
#1, #4 and #7	3.21 ± 11.82	0.43 ± 9.53	1.78 ± 5.47
#2, #5 and #8	-5.19 ± 11.41	4.39 ± 14.38	6.91 ± 7.05
#3, #6 and #9	-2.46 ± 9.99	-4.97 ± 9.39	4.83 ± 6.62

By computing the differentiation of the coordinates of the tip points, we were able to obtain the linear velocity of the actuator tip, which characterizes the dynamic behavior of the soft actuator. Figure 12 illustrates the predicted tip velocities in the x-, y-, and z-axis directions compared to the ground truth, demonstrating the ability of our method to capture the dynamic behavior of soft actuators. The RMSE values for the predicted tip speeds in the x-, y-, and z-axis directions were found to be 3.14 mm/s, 3.17 mm/s, and 1.44 mm/s, respectively. To evaluate the accuracy of the speed prediction, we computed the mean and standard deviation of the prediction errors, which are reported in Table 3.

Table 3. Prediction accuracy of actuator tip speed for validation data set.

Direction	Prediction Accuracy (mm/s)
x-axis	0.48 ± 3.11
y-axis	-0.25 ± 3.16
z-axis	0.11 ± 1.44

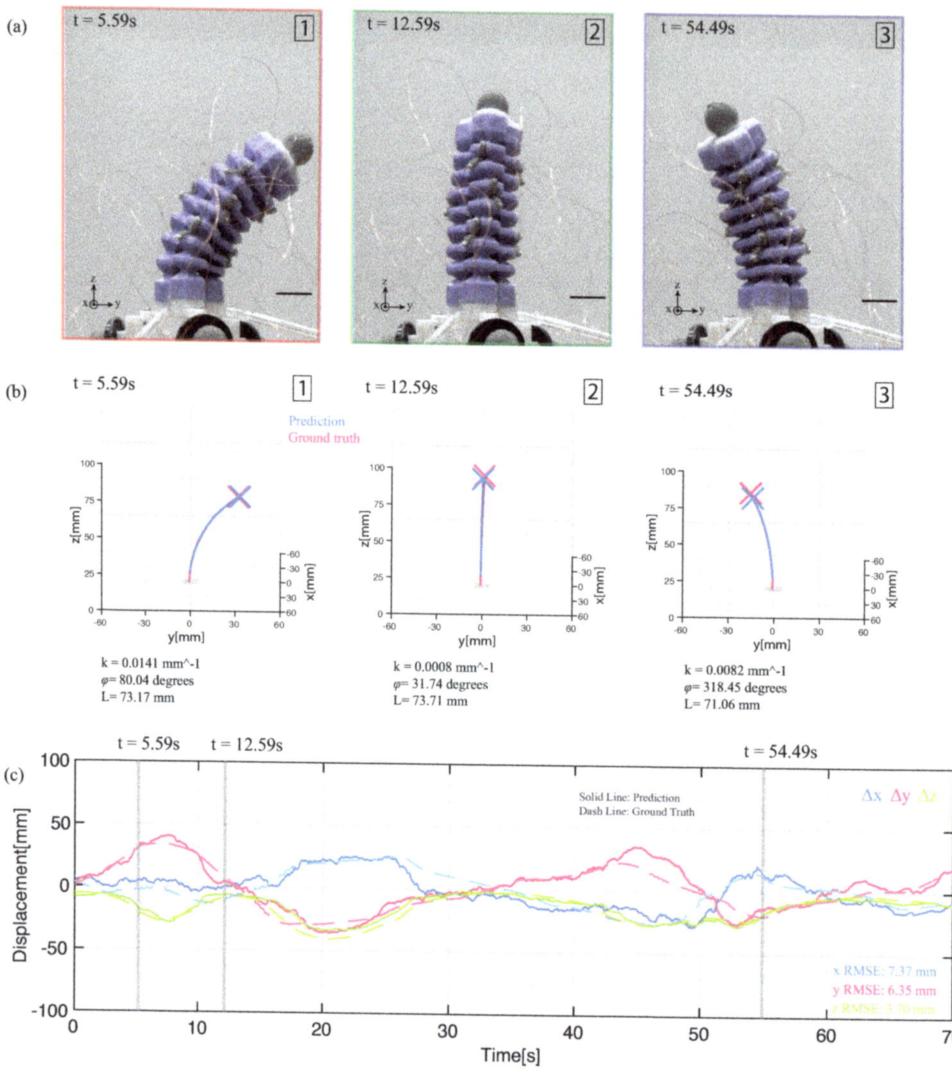

Figure 11. Validation results on random actuator motions. (**a**) Photos with time step. (**b**) Respective 3D representations of postures of the soft actuator. Kinematic parameters k, ϕ, and L are shown in the figure. In (**b**), the blue curve represents predictive posture; the red curve represents ground truth. (**c**) Plots of ground truth (dashed lines) and predictive displacement in the x−axis direction (blue curve), y−axis direction (red curve), and z−axis direction (green curve) of the reference frame. Dashed lines are representing ground truth. Time steps in (**a**,**b**) are represented by a gray shaded vertical bar (scale bar = 1 cm. The perspectives of (**b**) are top−down view along the x−z plane, parallel to the vector (3, 0, 1)). Actuator motions in (**c**) are included in Video S1.

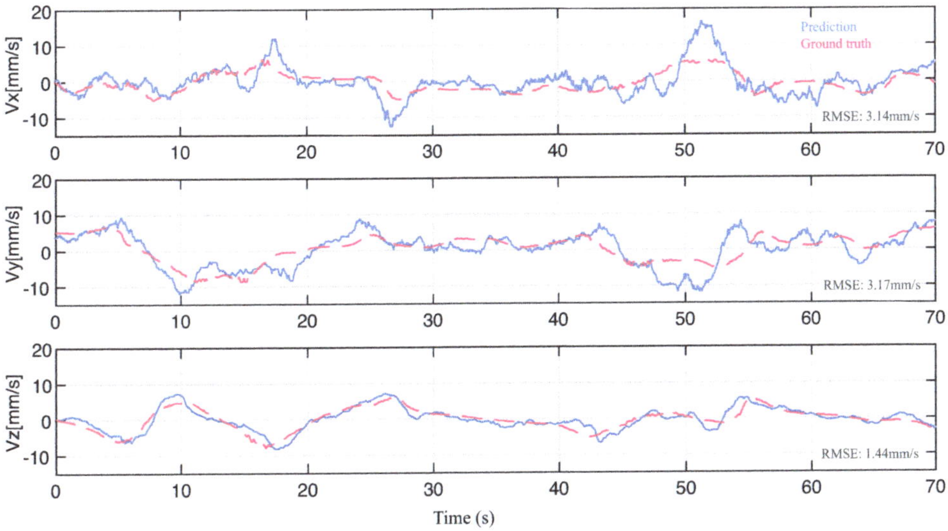

Figure 12. Velocity prediction of actuator tip for validation results in Figure 11c. From top to bottom, velocity in the prime axis direction of the reference frame was represented. Blue curves represent predictive velocity; red dash lines represent ground truth.

5. Conclusions and Discussion

This paper presents a novel end-to-end method for perceiving the 3D posture and dynamic behaviors of bellow-shaped soft actuators using distributed miniature flexible porous piezoresistive sensors. Additionally, we demonstrate the calibration of non-linear, hysteresis sensors using a trained RNN neural network. The scalable design of the sensor can be achieved by selecting a conductive sponge sheet with different thicknesses and laser-cutting patterns, making it applicable to different scales of soft robots. The porous structure and flexible material of the sensor minimize the influence on actuator motions, providing an advantage over previously developed flexible sensors, particularly in small-scale applications. A comparison between our flexible sponge sensor and recent typical works where RNN was applied in flexible sensor calibration is presented in Table 4. In comparison with the sensors developed in [38,46,47], we separated the design and fabrication of sensors from that of the attached actuator, thereby reducing fabrication difficulties and increasing design flexibility. Our flexible sponge sensor outperforms the sensors developed in [19] in capturing the velocity information due to the low inertia of the sponge material. Furthermore, compared with our previous work in [36,39], we have increased the degree-of-freedom of the motion.

5.1. Discussion on Kinematic Description of Soft Actuator

Although the posture of the soft actuator in this study was estimated based on the constant curvature concept, our end-to-end posture-perceiving method can be expanded to non-constant curvature situations. For soft actuators, this method is also suitable for estimating posture based on piecewise constant curvature (PCC), Bezier curves, or splines. The number and position of the sensor depend on the accuracy of the single sensor and the model employed to estimate the actuator posture, respectively. In this study, redundant sensors (theoretically only one sensor on each sub-chamber is necessary) were employed to improve the overall accuracy. For the piecewise constant curvature situation, each divided segment can be treated as having constant curvature, and the procedures in this study can be followed. For Bezier curves/splines situations, the sensor can be combined with the Bezier/spline model that enables non-constant curvature measurement. The neural network-based method can be used to find the parameters in the models or together with the model. The joint angle can be measured based on the method in this study, after which the joint angle distribution can be estimated. Finally, the overall posture of the soft actuator can be estimated.

Table 4. Comparison among our work and recent typical works where RNN was applied in flexible sensor calibration.

	Research Work	Sensor Material /Structure	Design Method	Advantages	Drawbacks
#1	Truby et al. [19]	Conductive Silicon	Separated	3-DoF Posture Estimation	Only Steady-State Estimation
#2	Shu et al. [39]		Separated	Dynamic Posture Estimation	Only Performs on Single DoF Motion
#3	Thuruthel et al. [38]		Integrated	2-DoF Posture and Contact Force Estimation	Low Sampling Frequency (10 Hz)
#4	Truby et al. [46]	Sealed Air Chamber	Integrated	3-DoF Dynamic Posture Estimation and Tactile Detection	Require Specific Fabrication Equipment (High-Resolution DLP 3D Printers)
#5	Ang et al. [47]		Integrated	2-DoF Dynamic Posture and Contact Force Estimation	Low Sampling Frequency
#7	Shu et al. (This Work)	Conductive Sponge	Separated	3-DoF Dynamic Posture Estimation	Saturation Region Affects Estimation Accuracy

5.2. Discussion on Sensor Characterization and Performance of the Sensing Method

As discussed in Section 4.1.2, both the sensitivity and resistive torque of the flexible sponge sensor exhibit a negative correlation with the angle β. Based on the results of ten diverse cycles of testing, it was observed that the electrical resistance of the flexible sponge sensor exhibited variations across different cycles. This issue poses a challenge for conventional model-based sensor calibration methods. However, the LSTM calibration neural network proposed in this paper was found to effectively address this challenge, enabling accurate calibration of the sensor and estimation of the posture of the soft actuators. It is worth noting that a saturation region arises when the sensor is consistently pressured. Specifically, when the sensors are adhered to the ravine of the bellow, the saturation occurs when the bellow is fully compressed by negative pressure. In the context of the present study, this saturation is reflected in the 3-DoF actuator under investigation when the actuator is fully compressed (i.e., all three bellows are fully compressed) or the bending angle θ is substantial (i.e., one or two bellows are fully compressed). Consequently, under these two conditions, the estimation error of the sensing system would be more pronounced, as evidenced by the data at 45 s when the bellow actuator is in random motion (see Figure 11). It is worth noting that this issue cannot be resolved by employing a calibration neural network, as there is an absence of effective training data.

The performance of velocity prediction is closely linked to the quality of the base material of the sensor, particularly its bandwidth with respect to mechanical vibration. With a larger bandwidth, the sensors can react to faster deformations of the attached objects. In the present study, we used an off-the-shelf conductive sponge whose base material is PU sponge, which has a slow recovery time to its original shape when pressure is removed. Consequently, when tracking the velocity of the actuator, the estimation error is substantial when the acceleration is large (reflected in Figure 12). Fortunately, for soft actuators, the response is typically slow.

During testing, it was observed that the flexible sponge sensor may experience fatigue after prolonged use, which is possibly due to damage of the conductive coatings. In the fabrication process of the conductive PU sponge, the base PU sponge was merged with a conductive solution. After drying, the internal fibers of the sponge were enveloped by a layer of conductive coating. During the use of the sensors, the coating may come into contact with neighboring coatings, leading to damage after prolonged use. Another issue pertains to the operating temperature, as the PU material is sensitive to high-temperature environments. The heat generated by friction during operation may also shorten the sensor's lifetime. The presence of noise resulting from bare enameled wire during sensor resistance measurement may also have an impact on the predictive accuracy of the calibration neural network.

Moving forward, we aim to enhance the sensitivity, accuracy, and longevity of the sensor. Moreover, we plan to investigate additional applications of the proposed sensing system within the realm of soft robotics.

Supplementary Materials: The following supporting information can be downloaded at: https://www.mdpi.com/article/10.3390/s23136189/s1, Video S1: Actuator Motions in Figure 10c.

Author Contributions: Conceptualization, J.S., J.W. and R.K.-y.T.; methodology, J.S. and J.W., K.C.-C.C., L.-F.Y. and R.K.-y.T.; validation, K.C.-C.C., L.-F.Y. and R.K.-y.T.; writing—original draft preparation, J.S. and J.W.; writing—review and editing, J.S., J.W. and R.K.-y.T.; supervision, R.K.-y.T. and Z.L.; funding acquisition, R.K.-y.T. and Z.L. All authors have read and agreed to the published version of the manuscript.

Funding: The research is co-funded by the Guangdong Science and Technology Research Council (Ref. No. 2020B1515120064) and the Innovation and Technology Fund, HKSAR (Ref. No. ITS/156/20FP).

Institutional Review Board Statement: Not applicable.

Informed Consent Statement: Not applicable.

Data Availability Statement: Not applicable.

Acknowledgments: The authors would like to thank the Chinese University of Hong Kong for providing us with the opportunity to work in soft robotics research.

Conflicts of Interest: The authors declare no conflict of interest.

Abbreviations

The following abbreviations are used in this manuscript:

DOF	Degree of Freedom
PCC	Piecewise Constant Curvature
CC	Constant Curvature
LSTM	Long Short-Term Memory
RNN	Recurrent Neural Network
SD	Standard Deviation

Appendix A. Experimental Setup for Sensor Characterizing

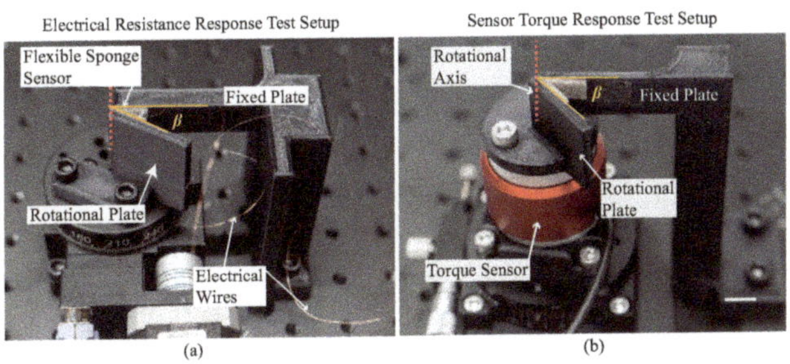

Figure A1. Experimental setup for sensor (**a**) electrical resistance response test and (**b**) sensor torque test (scale bar = 1 cm).

References

1. Rus, D.; Tolley, M.T. Design, fabrication and control of soft robots. *Nature* **2015**, *521*, 467–475. [CrossRef] [PubMed]
2. Pan, M.; Yuan, C.; Liang, X.; Dong, T.; Liu, T.; Zhang, J.; Zou, J.; Yang, H.; Bowen, C. Soft actuators and robotic devices for rehabilitation and assistance. *Adv. Intell. Syst.* **2022**, *4*, 2100140. [CrossRef]
3. Tang, Z.Q.; Heung, H.L.; Tong, K.Y.; Li, Z. Model-based online learning and adaptive control for a "human-wearable soft robot" integrated system. *Int. J. Robot. Res.* **2021**, *40*, 256–276. [CrossRef]
4. Yang, Y.; Chen, Y.; Li, Y.; Wang, Z.; Li, Y. Novel variable-stiffness robotic fingers with built-in position feedback. *Soft Robot.* **2017**, *4*, 338–352. [CrossRef] [PubMed]
5. Yuen, M.C.; Tonoyan, H.; White, E.L.; Telleria, M.; Kramer, R.K. Fabric sensory sleeves for soft robot state estimation. In Proceedings of the 2017 IEEE international conference on robotics and automation (ICRA), Singapore, 29 May–3 June 2017; pp. 5511–5518.
6. Gillespie, M.T.; Best, C.M.; Killpack, M.D. Simultaneous position and stiffness control for an inflatable soft robot. In Proceedings of the 2016 IEEE international conference on robotics and automation (ICRA), Stockholm, Sweden, 16–21 May 2016; pp. 1095–1101.
7. Larson, C.; Peele, B.; Li, S.; Robinson, S.; Totaro, M.; Beccai, L.; Mazzolai, B.; Shepherd, R. Highly stretchable electroluminescent skin for optical signaling and tactile sensing. *Science* **2016**, *351*, 1071–1074. [CrossRef]
8. Zhao, H.; O'Brien, K.; Li, S.; Shepherd, R.F. Optoelectronically innervated soft prosthetic hand via stretchable optical waveguides. *Sci. Robot.* **2016**, *1*, eaai7529. [CrossRef]
9. Wall, V.; Zöller, G.; Brock, O. A method for sensorizing soft actuators and its application to the RBO hand 2. In Proceedings of the 2017 IEEE International Conference on Robotics and Automation (ICRA), Singapore, 29 May–3 June 2017; pp. 4965–4970.
10. Truby, R.L.; Katzschmann, R.K.; Lewis, J.A.; Rus, D. Soft robotic fingers with embedded ionogel sensors and discrete actuation modes for somatosensitive manipulation. In Proceedings of the 2019 2nd IEEE International Conference on Soft Robotics (RoboSoft), Seoul, Republic of Korea, 14–18 April 2019; pp. 322–329.
11. Li, Y.; Guo, C.; Xin, W.; Pan, T.; Li, W.; Chiu, P.W.Y.; Li, Z. Design and Preliminary Evaluation of an Electromagnetically Actuated Soft-Tethered Colonoscope. *IEEE Trans. Med. Robot. Bionics* **2021**, *2*, 402–413. [CrossRef]
12. Elkeran, M.; Fanni, M. A transient FEA-based methodology for designing soft surgical manipulators. In Proceedings of the 2018 IEEE International Conference on Mechatronics and Automation (ICMA), Changchun, China, 5–8 August 2018; pp. 568–573.
13. Katzschmann, R.K.; Della Santina, C.; Toshimitsu, Y.; Bicchi, A.; Rus, D. Dynamic motion control of multi-segment soft robots using piecewise constant curvature matched with an augmented rigid body model. In Proceedings of the 2019 2nd IEEE International Conference on Soft Robotics (RoboSoft), Seoul, Republic of Korea, 14–18 April 2019; pp. 454–461.
14. Bajo, A.; Goldman, R.E.; Simaan, N. Configuration and joint feedback for enhanced performance of multi-segment continuum robots. In Proceedings of the 2011 IEEE International Conference on Robotics and Automation (ICRA), Shanghai, China, 9–13 May 2011; pp. 2905–2912.
15. Zhang, B.; Fan, Y.; Yang, P.; Cao, T.; Liao, H. Worm-like soft robot for complicated tubular environments. *Soft Robot.* **2019**, *6*, 399–413. [CrossRef]
16. Polygerinos, P.; Wang, Z.; Galloway, K.C.; Wood, R.J.; Walsh, C.J. Soft robotic glove for combined assistance and at-home rehabilitation. *Robot. Auton. Syst.* **2015**, *73*, 135–143. [CrossRef]
17. Yan, X.; Bowen, C.R.; Yuan, C.; Hao, Z.; Pan, M. Carbon fibre based flexible piezoresistive composites to empower inherent sensing capabilities for soft actuators. *Soft Matter* **2019**, *15*, 8001–8011. [CrossRef]
18. Pan, M.; Yuan, C.; Anpalagan, H.; Plummer, A.; Zou, J.; Zhang, J.; Bowen, C. Soft Controllable Carbon Fibre-based Piezoresistive Self-Sensing Actuators. *Actuators* **2020**, *9*, 79. [CrossRef]
19. Truby, R.L.; Santina, C.D.; Rus, D. Distributed Proprioception of 3D Configuration in Soft, Sensorized Robots via Deep Learning. *IEEE Robot. Autom. Lett.* **2020**, *5*, 3299–3306. [CrossRef]
20. Ye, Z.; Pang, G.; Liang, Y.; Lv, H.; Xu, K.; Wu, H.; Yang, G. Highly Stretchable and Sensitive Strain Sensor Based on Porous Materials and Rhombic-Mesh Structures for Robot Teleoperation. *Adv. Sens. Res.* **2023**, 2300044. [CrossRef]
21. Xu, K.; Lu, Y.; Takei, K. Flexible hybrid sensor systems with feedback functions. *Adv. Funct. Mater.* **2021**, *31*, 2007436. [CrossRef]
22. Kong, D.; Yang, G.; Pang, G.; Ye, Z.; Lv, H.; Yu, Z.; Wang, F.; Wang, X.V.; Xu, K.; Yang, H. Bioinspired Co-Design of Tactile Sensor and Deep Learning Algorithm for Human–Robot Interaction. *Adv. Intell. Syst.* **2022**, *4*, 2200050. [CrossRef]
23. Xu, K.; Fujita, Y.; Lu, Y.; Honda, S.; Shiomi, M.; Arie, T.; Akita, S.; Takei, K. A wearable body condition sensor system with wireless feedback alarm functions. *Adv. Mater.* **2021**, *33*, 2008701. [CrossRef]
24. Heng, W.; Yang, G.; Kim, W.S.; Xu, K. Emerging wearable flexible sensors for sweat analysis. *Bio-Des. Manuf.* **2022**, *5*, 64 –84. [CrossRef]
25. Xu, K.; Lu, Y.; Honda, S.; Arie, T.; Akita, S.; Takei, K. Highly stable kirigami-structured stretchable strain sensors for perdurable wearable electronics. *J. Mater. Chem. C* **2019**, *7*, 9609–9617. [CrossRef]
26. Vella, D. Buffering by buckling as a route for elastic deformation. *Nat. Rev. Phys.* **2019**, *1*, 425–436. [CrossRef]
27. Lo, L.W.; Zhao, J.; Wan, H.; Wang, Y.; Chakrabartty, S.; Wang, C. A soft sponge sensor for multimodal sensing and distinguishing of pressure, strain, and temperature. *ACS Appl. Mater. Interfaces* **2022**, *14*, 9570–9578. [CrossRef]
28. Li, X.; Li, X.; Liu, T.; Lu, Y.; Shang, C.; Ding, X.; Zhang, J.; Feng, Y.; Xu, F.J. Wearable, washable, and highly sensitive piezoresistive pressure sensor based on a 3D sponge network for real-time monitoring human body activities. *ACS Appl. Mater. Interfaces* **2021**, *13*, 46848–46857. [CrossRef] [PubMed]

29. Wei, Q.; Chen, G.; Pan, H.; Ye, Z.; Au, C.; Chen, C.; Zhao, X.; Zhou, Y.; Xiao, X.; Tai, H.; et al. MXene-sponge based high-performance piezoresistive sensor for wearable biomonitoring and real-time tactile sensing. *Small Methods* **2022**, *6*, 2101051. [CrossRef] [PubMed]
30. Li, Y.; Cui, Y.; Zhang, M.; Li, X.; Li, R.; Si, W.; Sun, Q.; Yu, L.; Huang, C. Ultrasensitive pressure sensor sponge using liquid metal modulated nitrogen-doped graphene nanosheets. *Nano Lett.* **2022**, *22*, 2817–2825. [CrossRef] [PubMed]
31. Yang, W.; Xie, M.; Zhang, X.; Sun, X.; Zhou, C.; Chang, Y.; Zhang, H.; Duan, X. Multifunctional soft robotic finger based on a nanoscale flexible temperature–pressure tactile sensor for material recognition. *ACS Appl. Mater. Interfaces* **2021**, *13*, 55756–55765. [CrossRef] [PubMed]
32. Chen, H.; Su, Z.; Song, Y.; Cheng, X.; Chen, X.; Meng, B.; Song, Z.; Chen, D.; Zhang, H. Omnidirectional bending and pressure sensor based on stretchable CNT-PU sponge. *Adv. Funct. Mater.* **2017**, *27*, 1604434. [CrossRef]
33. Ge, G.; Cai, Y.; Dong, Q.; Zhang, Y.; Shao, J.; Huang, W.; Dong, X. A flexible pressure sensor based on rGO/polyaniline wrapped sponge with tunable sensitivity for human motion detection. *Nanoscale* **2018**, *10*, 10033–10040. [CrossRef] [PubMed]
34. Ye, Z.; Pang, G.; Xu, K.; Hou, Z.; Lv, H.; Shen, Y.; Yang, G. Soft robot skin with conformal adaptability for on-body tactile perception of collaborative robots. *IEEE Robot. Autom. Lett.* **2022**, *7*, 5127–5134. [CrossRef]
35. Ye, Z.; Yang, G.; Pang, G.; Huang, X.; Yang, H. Design and implementation of robot skin using highly sensitive sponge sensor. *IEEE Trans. Med. Robot. Bionics* **2020**, *2*, 670–680. [CrossRef]
36. Shu, J.; Wang, J.; Lau, S.C.Y.; Su, Y.; Heung, K.H.L.; Shi, X.; Li, Z.; Tong, R.K.y. Soft Robots' Dynamic Posture Perception Using Kirigami-Inspired Flexible Sensors with Porous Structures and Long Short-Term Memory (LSTM) Neural Networks. *Sensors* **2022**, *22*, 7705. [CrossRef]
37. Elgeneidy, K.; Lohse, N.; Jackson, M. Bending angle prediction and control of soft pneumatic actuators with embedded flex sensors—A data-driven approach. *Mechatronics* **2018**, *50*, 234–247. [CrossRef]
38. Thuruthel, T.G.; Shih, B.; Laschi, C.; Tolley, M.T. Soft robot perception using embedded soft sensors and recurrent neural networks. *Sci. Robot.* **2019**, *4*, eaav1488. [CrossRef] [PubMed]
39. Shu, J.; Wang, J.; Su, Y.; Liu, H.; Li, Z.; Tong, R.K.Y. An End-to-end Posture Perception Method for Soft Bending Actuators Based on Kirigami-inspired Piezoresistive Sensors. In Proceedings of the 2022 IEEE-EMBS International Conference on Wearable and Implantable Body Sensor Networks (BSN), Ioannina, Greece, 27–30 September 2022; pp. 1–5.
40. Hochreiter, S.; Schmidhuber, J. Long short-term memory. *Neural Comput.* **1997**, *9*, 1735–1780. [CrossRef] [PubMed]
41. Shih, B.; Drotman, D.; Christianson, C.; Huo, Z.; White, R.; Christensen, H.I.; Tolley, M.T. Custom soft robotic gripper sensor skins for haptic object visualization. In Proceedings of the 2017 IEEE/RSJ international conference on intelligent robots and systems (IROS), Vancouver, BC, Canada, 24–28 September 2017; pp. 494–501.
42. Galloway, K.C.; Chen, Y.; Templeton, E.; Rife, B.; Godage, I.S.; Barth, E.J. Fiber optic shape sensing for soft robotics. *Soft Robot.* **2019**, *6*, 671–684. [CrossRef]
43. Shu, J.; Chirarattananon, P. A quadrotor with an origami-inspired protective mechanism. *IEEE Robot. Autom. Lett.* **2019**, *4*, 3820–3827. [CrossRef]
44. Webster, R.J., III; Jones, B.A. Design and kinematic modeling of constant curvature continuum robots: A review. *Int. J. Robot. Res.* **2010**, *29*, 1661–1683. [CrossRef]
45. Kingma, D.P.; Ba, J. Adam: A method for stochastic optimization. *arXiv* **2014**, arXiv:1412.6980.
46. Truby, R.L.; Chin, L.; Zhang, A.; Rus, D. Fluidic innervation sensorizes structures from a single build material. *Sci. Adv.* **2022**, *8*, eabq4385. [CrossRef]
47. Ang, B.W.K.; Yeow, C.H. A learning-based approach to sensorize soft robots. *Soft Robot.* **2022**, *9*, 1144–1153. [CrossRef]

sensors

MDPI

Article

Optimizing Appearance-Based Localization with Catadioptric Cameras: Small-Footprint Models for Real-Time Inference on Edge Devices

Marta Rostkowska * and Piotr Skrzypczyński

Institute of Robotics and Machine Intelligence, Poznan University of Technology, 60-965 Poznan, Poland; piotr.skrzypczynski@put.poznan.pl
* Correspondence: martarostkowska89@gmail.com

Abstract: This paper considers the task of appearance-based localization: visual place recognition from omnidirectional images obtained from catadioptric cameras. The focus is on designing an efficient neural network architecture that accurately and reliably recognizes indoor scenes on distorted images from a catadioptric camera, even in self-similar environments with few discernible features. As the target application is the global localization of a low-cost service mobile robot, the proposed solutions are optimized toward being small-footprint models that provide real-time inference on edge devices, such as Nvidia Jetson. We compare several design choices for the neural network-based architecture of the localization system and then demonstrate that the best results are achieved with embeddings (global descriptors) yielded by exploiting transfer learning and fine tuning on a limited number of catadioptric images. We test our solutions on two small-scale datasets collected using different catadioptric cameras in the same office building. Next, we compare the performance of our system to state-of-the-art visual place recognition systems on the publicly available COLD Freiburg and Saarbrücken datasets that contain images collected under different lighting conditions. Our system compares favourably to the competitors both in terms of the accuracy of place recognition and the inference time, providing a cost- and energy-efficient means of appearance-based localization for an indoor service robot.

Keywords: omnidirectional vision; mobile robot; localization; deep learning; edge computing

Citation: Rostkowska, M.; Skrzypczyński, P. Optimizing Appearance-Based Localization with Catadioptric Cameras: Small-Footprint Models for Real-Time Inference on Edge Devices. *Sensors* **2023**, *23*, 6485. https://doi.org/10.3390/s23146485

Academic Editors: Xin Zhao, Mingzhu Sun and Qili Zhao

Received: 18 June 2023
Revised: 14 July 2023
Accepted: 15 July 2023
Published: 18 July 2023

1. Introduction

The rapid development of robotics and artificial intelligence applications is leading to the proliferation of mobile service robots [1,2]. Technological advancements, such as artificial intelligence and machine learning, have significantly improved the capabilities and autonomy of these robots, making them more efficient and reliable in performing various tasks. Additionally, the increasing demand for automation and efficiency in industries such as healthcare, hospitality, and logistics has created a strong market incentive for developing and deploying service mobile robots.

Also, the growing need for eldercare robots has become increasingly evident as the global population ages. These robots can provide valuable assistance and companionship to older adults, monitoring their health and enhancing their overall well-being [3]. However, these robots must be affordable to ensure widespread accessibility and adoption among families and caregivers [4].

A common requirement in these service robots is to be able to localize within their workspace, which is usually a man-made indoor environment [5]. Although precise position tracking can be provided by a SLAM (simultaneous localization and mapping) system using vision or RGB-D data, the issue of global localization remains a problem when the robot's previous position data cannot be used [6]. Such a problem in practice arises, for example, in dynamic environments due to occlusions. There are practical global localization algorithms, such as the one proposed in our previous work [7], but they have two functional limitations, namely, they require long-range sensors to extract features that are distant from the robot and are computationally expensive. These features make them unsuitable for a small and inexpensive service robot.

Therefore, we propose a solution to the problem of global localization in a known (entirely or partially) environment using a passive catadioptric camera and the principle of recognizing places previously visited by the robot (Figure 1). The applied sensor with a catadioptric camera is a variant of the biologically inspired sensor with a hybrid field of view that we introduced in [8]. This sensor uses a catadioptric camera to achieve omnidirectional vision, an analogue of the peripheral vision found in vertebrates [9]. It allows animals to orient themselves to changes and hazards in the environment quickly. The sensor described in [8,9] is complemented by a moving perspective camera that performs the functions of foveal vision, the more accurate but spatially limited vision mode in animals. This function is not used in the research presented in this article, as we limit the scope to global appearance-based localization, i.e., the assignment of the robot's current location to one of the previously recognized (visited) places. Our approach yields information about the similarity of the places observed in the current perception and locations stored in a reference map.

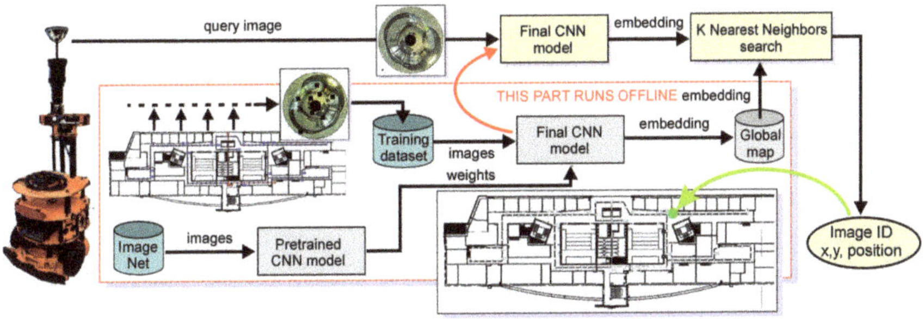

Figure 1. Overview of the proposed method—a flowchart of the appearance-based localization system. The service robot is shown with the latest, larger-field-of-view catadioptric camera, but without the perspective camera, which is not used in this research.

Although appearance-based localization does not provide an accurate metric position of the robot in a global reference system, the ability to tell if the robot is close to one of the known locations is often sufficient for indoor navigation [10]. If a map of reference places is collected at high density (e.g., based on a grid with cells of one meter in size or smaller), this kind of localization may be sufficient for the service robot's tasks. In addition, appearance-based localization can be supplemented by visual odometry or the recognition of artificial landmarks deployed at a given location [11]. The perspective camera of a hybrid sensor can be used to perform these functions. The main objectives of this research work are the following:

- Experimental analysis of neural network architectures in search of an architecture for an image-based place recognition system suitable for implementation on an embedded computer of an intelligent vision sensor with limited power and resources.
- Experimental verification of the possibility of using catadioptric camera images in the appearance-based localization task without developing them into panoramic form significantly reduces the computational load.
- Analysis of the strategy for creating training sets in a place recognition task, assuming that the obtained solution should be generalized to different image acquisition conditions, mainly depending on illumination.

We propose a novel approach that adopts a convolutional neural network (CNN) architecture to directly process the omnidirectional images for real-time place recognition to meet these objectives. CNNs are specialized for processing grid-like data, particularly images, using convolutional layers and parameter sharing to capture spatial patterns effectively. The proposed system leverages the concept of global image descriptors, which are already proven to be efficient in place recognition [12]. We employ a CNN to produce the descriptors in the form of embedding vectors directly from the omnidirectional images, thus avoiding the processing overhead required for computing undistorted panoramic images, which are often used in appearance-based localization with catadioptric cameras [13]. The proposed architecture is optimized for inference on the Nvidia Jetson TX2 edge computing platform integrated with our sensor. The low-cost Jetson TX2 board is designed for peak processing efficiency at only 7.5 W power. Regarding energy consumption for image processing, the Jetson TX2 has a clear advantage over an x86-based platform [14]. While the exact power consumption will depend on the specific image processing workload, the Jetson TX2 is designed to provide a good balance between performance and energy efficiency [15]. Hence, by applying an integrated sensor with an edge computing platform and developing a matching small-footprint neural network architecture, we obtain a self-contained, energy-efficient, and compact system for real-time appearance-based localization that can be integrated with practically any mobile service robot, providing this robot with reliable global localization capabilities at low cost. The contribution of this paper is threefold:

1. A novel, simple-yet-efficient CNN-based architecture of the appearance-based localization system that leverages a lightweight CNN backbone trained to apply transfer learning to produce the embeddings and the K-nearest neighbours method for quickly finding an embedding matching the current perception.
2. A thorough experimental investigation of this architecture, considering several backbone network candidates and omnidirectional or panoramic images used to produce the embeddings. The experiments were conducted on three different datasets: two collected with variants of our bioinspired sensor and one publicly available.
3. An investigation of the strategies for creating the training set and the reference map for the localization system conducted on the COLD Freiburg dataset. This part of our research allowed us to test how our neural network model generalizes to images acquired under different lighting/weather conditions. It resulted in the recommendation of using data balanced concerning their acquisition parameters, improving generalization.

The remainder of this article is structured as follows. Most important related works are reviewed in Section 2. Section 3 introduces the proposed architecture of the localization system and details the neural networks being used. Next, Section 4 describes the experimental setups and dataset used to test various aspects of the proposed solution, while Section 5 provides the results of experiments and contributes an in-depth analysis of the performance of different variants of the investigated system. Finally, Section 6 concludes the article and proposes future extensions.

2. Related Work

Appearance-based localization from omnidirectional images has garnered significant attention in computer vision and robotics. Researchers have developed various techniques to address the challenges posed by the distortion and wide field of view of omnidirectional cameras. This section reviews the most relevant works that have contributed to the state of the art in this area.

The application of passive vision sensors for localization was extensively researched in robotics, resulting in several visual Simultaneous Localization and Mapping (SLAM) algorithms [16]. However, the applications of visual SLAM on commercially viable mobile robots are limited by the often-insufficient on-board computing resources of such robotic platforms and due to problems raised by the changing lighting conditions, rapid changes of viewpoint while the robot is moving, and the lack of salient local features in some indoor environments. Moreover, SLAM does not guarantee to solve the global localization problem whenever the robot loses track of its pose due to any of the issues mentioned above [17].

Therefore, the appearance-based recognition of locations becomes an exciting addition to visual SLAM for closing the loops and relocalizing a lost robot [18]. This approach, in many variants, is also considered a localization method on its own, which is particularly suitable for large-scale outdoor scenarios [10]. Unlike the visual SLAM algorithms, appearance-based localization methods only determine if the observed scene resembles an already visited location. However, the place recognition methods scale better for large environments than typical SLAM algorithms [19]. In this context, catadioptric cameras yielding omnidirectional images improve the reliability of place recognition for robot localization in comparison to the narrow-field-of-view perspective cameras, as demonstrated by the work on the COsy Localization Database (COLD) dataset [20], which we also use to evaluate our localization system. An interesting research direction is to use image sequences instead of individual images, which decreases the number of false positives in place recognition for environments with self-similarities and increases the robustness of scene dynamics [12]. We applied this idea in our earlier work on place recognition for mobile devices [21], making it possible to implement robust place recognition on a smartphone with very limited computing power, while still using nondistorted perspective images.

In the appearance-based methods, each image is described by descriptors of salient features contained in this image, or is directly described by a whole-image descriptor. Although SURF features were used directly in appearance-based localization performing image retrieval in a hierarchical approach [22], the direct matching of local features is considered inefficient for place recognition [10] if point feature descriptors are used (such as the popular SIFT, SURF, and ORB [23]). Hence, the bag of visual words (BoVW) technique [24] is commonly used, which organizes the features into a visual vocabulary. Next, images described by visual words can be efficiently matched by comparing binary strings or histograms. One prominent example of a location recognition algorithm employing the BoVW technique is FAB-MAP [25,26], which efficiently compares images with a histogram-based approach.

Global image descriptors have proven effective for capturing the overall appearance of omnidirectional images [27]. Earlier works focused on adapting existing, general-purpose feature extraction and matching algorithms. Menegatti et al. [28] proposed using the Fourier transform to handle geometric distortions in catadioptric images. More recently, Payá et al. [29] introduced a method based on the Radon transform to extract global environmental descriptions from omnidirectional images. These works provided foundations for subsequent research by addressing the specific characteristics of omnidirectional images. Examples of hand-crafted descriptors adopted for the global description of omnidirectional images include HOG (histogram of oriented gradients) [30] and Gist [31], which were applied to omnidirectional images from a catadioptric camera in appearance-based localization by Cebollada et al. [32]. While both these methods of image description provided relatively efficient descriptions of the images, allowing the localization system to recognize the places accurately, the descriptor construction algorithms initially developed for perspective camera images required the catadioptric images to be undistorted and converted to panoramic images, which creates a significant computation overhead.

Machine learning methods have gained popularity in place recognition, also from omnidirectional images [33]. Working with typical perspective images, Li et al. [34] proposed an image similarity measurement method based on deep learning, which combines local and global features to describe the image and can be used for indoor place recognition for a robotic agent. Significant progress in appearance-based localization and navigation was achieved by the NetVLAD approach [35], a CNN-based method that aggregates local features for global image representation. The NetVLAD network consists of a CNN for feature extraction and a layer based on vector of locally aggregated descriptors—VLAD [36]. In this architecture, VLAD is a feature quantization technique similar in concept to the bag of visual words idea, as it captures information about the statistics of an image's local descriptors. The VLAD is a method for combining descriptors for both instance-level searches [37] and image classification [38]. Although Cheng et al. [39] used NetVLAD with panoramic images from an omnidirectional system, this approach was demonstrated successfully, mainly in outdoor scenarios working with perspective camera images. For indoor scenarios, [13] introduced the omnidirectional convolutional neural network (O-CNN) architecture, which, similarly to our approach, is trained to retrieve the closest place example from the map. Whereas the O-CNN architecture takes advantage of the omnidirectional view by incorporating circular padding and rotation invariance, it requires the omnidirectional images to be converted to their panoramic counterparts. Also, Cebollada et al. [40] demonstrated the benefits of solving localization problems as a batch image retrieval problem by comparing descriptors obtained from intermediate layers of a CNN. A CNN processing rectangular panoramic images reconstructed from the original catadioptric input is used in this work.

As the construction of invariant feature descriptors for omnidirectional images is problematic, Masci et al. [41] proposed to learn invariant descriptors with a similarity-preserving hashing framework and a neural network to solve the underlying optimization problem. Ballesta et al. [42] implemented hierarchical localization with omnidirectional images using a CNN trained to solve a classification task for distinguishing between different rooms in the environment and then a CNN trained for regression of the pose within the recognized room. Although this solution does not require converting the catadioptric images into panoramic ones, its performance is limited by the employed two-stage scheme with separated classification and regression steps. More recent work from the same team [43] solved the appearance-based localization problem by applying a hierarchical approach with the AlexNet CNN. Assuming an indoor environment, they first accomplished a room retrieval task and then carried out the fine localization step within the retrieved room. To this end, the CNN was trained to produce a descriptor, which was compared with the visual model of the selected room using a nearest neighbour search. This approach does not require panoramic conversion of the collected catadioptric images and is overall most similar to the solution proposed in this paper. However, we introduce a much simpler, single-stage architecture based on a recent, lightweight CNN backbone, and the concept of direct retrieval of the image stored in the environment map, which is most similar in appearance to the query image. The efficient process of constructing the embeddings from a pretrained CNN, followed by a fast comparison of these embeddings/descriptors in the KNN framework, allowed us to give up with separated room retrieval in favour of a single-stage architecture, which suits our embedded computing platform well. We compare it directly to the results shown in [43] on the COLD Freiburg dataset, demonstrating our approach's superior performance and real-time capabilities.

3. Localization System Architecture

In the proposed localization system, the robot figures out its current location by determining the similarity between the currently captured image (query image) and images stored in a database (map) describing the environment. This task refers to efficient, real-time image retrieval [10]. The localization procedure involves comparing a global descriptor constructed in real time from the image currently captured by the robot with a previously prepared database of descriptors representing the images of previously visited places and finding the image with the highest possible similarity in the feature space (Figure 2). Each location has its representation in the prepared database of images, and the locations where the images were taken are assumed to cover the entire robot's workspace. Images from the database are recorded at known locations, so finding one with the minimum distance (in the sense of similarity of appearance) to the current perception allows our robotic agent to approximate its location in the real world.

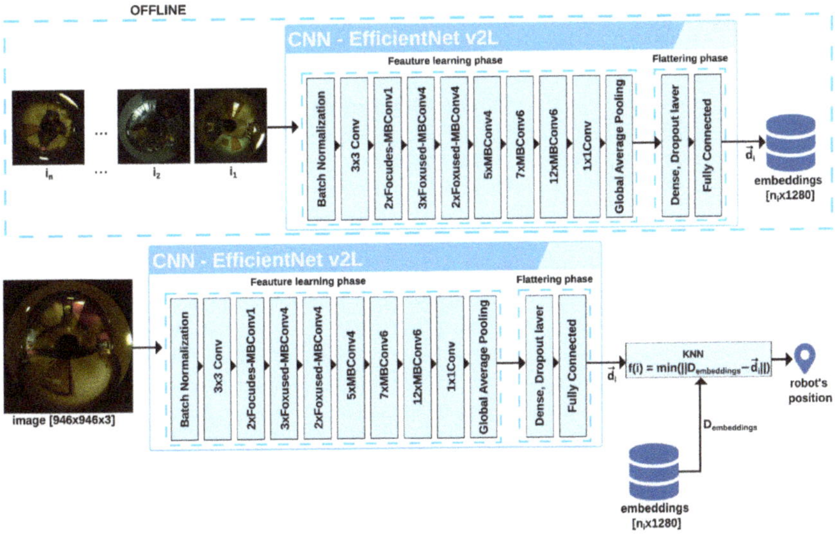

Figure 2. Diagram of the CNN-based image description blocks that produce embeddings used as global descriptors in the localization system. The global map is built from n_i images $(i_1 \dots i_{n_i})$ converted to embedding vectors $\vec{\mathbf{d}}_i$ that are stored in the map $\mathbf{D}_{\text{embeddings}}$ of n_i embeddings (global descriptors). Note that panoramic images can be used as well instead of the omnidirectional ones.

The proposed localization system uses a CNN to determine the set of natural features for a given location, and the K-nearest neighbours (KNN) [44,45] algorithm to find the closest image from the provided database of images.

CNN and KNN are both machine learning techniques, but differ in their approach and application. CNN learns hierarchical representations of data through multiple convolutional layers, pooling, and fully connected layers. In contrast, KNN is a simple and intuitive algorithm for classification and regression tasks. It makes predictions based on the similarity of new data points to the existing labelled data points in the feature space. One can use CNN to extract features from images and then apply KNN to those extracted features for classification. This hybrid approach leverages the strengths of both algorithms, with CNN capturing intricate patterns and KNN using the extracted features for classification [46]. This idea is used in our localization system. The backbone CNN creates descriptors, which hereinafter are also called "embeddings", directly from the omnidirectional images, avoiding the additional computations required to obtain undistorted panoramic images. The KNN algorithm uses the embeddings that encode the most salient features of the observed places, to find in the database (i.e., the global map) the images that best match the current observation. In Section 4, we demonstrate that the accuracy of localization with raw catadioptric images is at least as good as with the converted panoramic images, while it demands less computing power.

The preparation of the CNN model is based on training the network to correctly recognize places, with the specific aim of training the higher layers of the network to extract feature maps specific to each location properly. Because the CNN used as the backbone of our system is pretrained on images unrelated to the target domain (ImageNet dataset [47] was used in pretraining), the network was fine-tuned before use by unfreezing several layers and training on the target domain images using cross-entropy as a loss function. Cross-entropy defines the distance between two probability distributions according to the equation:

$$\mathrm{H}(\mathrm{P}(\mathrm{y}, 1-\mathrm{y}), \mathrm{P}(\mathrm{y}_{\text{pred}}, 1-\mathrm{y}_{\text{pred}})) = \mathrm{P}(\mathrm{y}) \log\Big(\mathrm{P}(\mathrm{y}_{\text{pred}})\Big) + \mathrm{P}(1-\mathrm{y}) \log\Big(\mathrm{P}(1-\mathrm{y}_{\text{pred}})\Big), \tag{1}$$

where y—the actual location; y_{pred}—the location obtained via a neural network; $P(y, 1 - y)$—the probability distribution of the actual location; $P(y_{pred}, 1 - y_{pred})$—the probability distribution of the location determined via a neural network (prediction).

At first, the images are processed by the trained convolutional neural network, from which the output layer was removed, to obtain descriptors (in the form of embedding vectors) that describe the global characteristic features of each image in the database, i.e., each unique place visited by the robot. In this way, a global map of all locations based on reference images is created. Not all convolutional network architectures from the literature can be used on a robotic onboard computer with fewer computational and memory resources.

This research uses backbone networks from the MobileNet [48] and EfficientNet [49] families, which are optimized for mobile devices while ensuring high accuracy with a minimal number of parameters and mathematical operations. The MobileNet model uses depth-separated convolution layers consisting of depth-wise convolution and point-wise convolution. Convolution concerning depth (spatial convolution) is used to apply a single filter for each input channel. In MobileNet V2, a new module with inverted residual structure has been introduced, there are two types of blocks. One is an inverted residual block of width 1. The other one is a block of width 2 to reduce the size of the feature map. There are three layers for both types of blocks. The first layer is a 1×1 convolution with the ReLU activation function, and the second layer is a convolution against depth. The third and final layer is another convolution of size 1×1, with linear bottlenecks. Residual blocks connect the beginning and end of the convolutional block via a skip connection. Adding these two states allows the network to access previous activations not modified in the convolution block. This approach has proven to be essential for building networks of large depths. In MobileNet V2, the basic convolutional layer is called MBConv and contains an inverted residual block with linear bottleneck and depth-separated convolution, with batch normalization behind each convolutional layer.

The EfficientNet model, which we have selected for our final architecture, can be seen as a further step towards efficiency compared to the MobileNet model. EfficientNet uses a complex model scaling technique based on a set of specified coefficients. Instead of randomly scaling width, depth, or resolution, compound scaling uniformly scales each dimension using some fixed scaling coefficient set. Such scaling only increases the predictive ability of the network by replicating the underlying convolutional operations and structure of the network. EfficientNet uses the MBConv blocks as in the MobileNet V2 network, but with a squeeze-and-excitation (SE—[50]) block being added. This structure helps reduce the overall number of operations required and the model's size.

The backbone CNN extracts from the image features that uniquely describe different locations and builds embedding vectors that serve as global image descriptors in our system. In the next step, the algorithm creates an index from the global map, which is used for efficient similarity search. The original images collected by the robot are no longer needed for localization and the obtained global map has a compact form. All operations to produce the global map are performed offline.

Then, to localize the robot, we need to query the global map (database of embeddings) with the descriptor/embedding produced from the current perception of the agent, which boils down to a similarity search task. Similarity search is a typical issue in machine learning solutions using embedding vectors, and becomes increasingly difficult as the vectors' dimensions and/or size increase. Classic methods for finding similarity between vector-described elements in an extensive database include linear search and search in K-D-trees [51]. K-D-trees are binary trees used to organize points representing data in a K-dimensional space and allow for a very efficient search of points in that space, including a nearest neighbour (NN) search, which we are interested in [52].

Each node in the tree represents a K-dimensional point. Each nonleaf node in the tree acts as a hyperplane, dividing the space into two parts. Using a K-D tree for nearest neighbour search involves finding the point in the tree that is closest to a given query point. For this purpose, the algorithm traverses the tree and compares the distance between the query point and points in each leaf node. Starting from the root node, it recursively moves down the tree until it reaches the leaf node, following the same procedure as when inserting a node. Many implementations of the nearest neighbour search using K-D-trees are known in Python, including the very popular SciKit-Learn library. However, for this project, we selected the Facebook AI Similarity Search (Faiss) library [53], written in C++ with wrappers for Python and support for GPU, which suits our implementation on Nvidia Jetson well. The Faiss library solves our similarity search problem using indexing and searching with the KNN method. Once the index type is selected, the algorithm processes the embedding vectors obtained from the neural network and places them in the index. The index can be stored on disk or in memory, and searching, adding, or removing items to the index can be performed in real-time. In addition, the Faiss library has an autotuning mechanism that scans the parameter space and selects those parameters that provide the best possible search time at a given accuracy.

Place recognition begins by loading the learned CNN model and index of images (map) into memory, and then the captured images (queries) are compared with the previously created image database using the KNN algorithm in the space of embedding vectors. The embeddings are compared using L2 (Euclidean) distance, which has been shown to be more computationally efficient than feature binarization followed by the comparison applying Hamming distance [36,54].

Once the similarity between the query image and the map is determined, the results are presented in the form of the image retrieval accuracy and the position error between the query image and the map image determined as the most similar one. As we assume that ground truth positions for all map images and query images are known, as in the COLD dataset [55], we simply use the Euclidean distance in metric space to quantify this error. The averaged Euclidean distance is used to calculate the position error over an entire experiment involving many queries. The arithmetic mean is calculated over all places according to the equation:

$$\overline{\overline{b_L}} = \frac{\sum_{i=1}^{n} \sqrt{(x_{gt_i} - x_{e_i})^2 + (y_{gt_i} - y_{e_i})^2}}{n}, \tag{2}$$

where $\overline{\overline{b_L}}$—the average position measurement error; n—the number of query images; x_{gt_i}—the x coordinate for the ground truth location of the *i*-th query image; x_{e_i}—the x coordinate for the estimated location of the *i*-th query image; y_{gt_i}—the y coordinate of the estimated location of the *i*-th query image; y_{e_i}—the y coordinate for the estimated location of the *i*-th query image.

The architecture of the localization system shown in its general form in Figure 1 was tested in several variants differing in the type of neural network used as an extractor of image embeddings and the use of catadioptric camera images directly or images converted to panoramic form. The suitability of the NetVLAD approach in the described system was also investigated. The investigated variants are described in the next section of this paper.

4. Experiments

To confirm the proposed solution's effectiveness and determine the best-performing CNN architecture, experiments were carried out at the Mechatronics Centre of Poznań University of Technology using two catadioptric cameras with different parameters. Then, experiments with the publicly available COLD database were carried out to demonstrate the performance of our approach with respect to selected state-of-the-art solutions in appearance-based localization on this dataset.

4.1. Experiment 1: Integrated Sensor on a Mobile Robot

A Labbot robot (Figure 3a) with an integrated catadioptric vision sensor was used in the first scenario. The catadioptric camera in this sensor consists of a Microsoft LifeCam and a hyperbolic mirror, which provides a field of view of 360° and produces images with a resolution of 640 × 480. The images are processed by a Nvidia Jetson TX2 computer integrated with the sensor [8] The Jetson TX2 offers a 256-core Pascal architecture General Purpose Graphics Processing Unit (GPGPU) to support the real-time operation of the localization system.

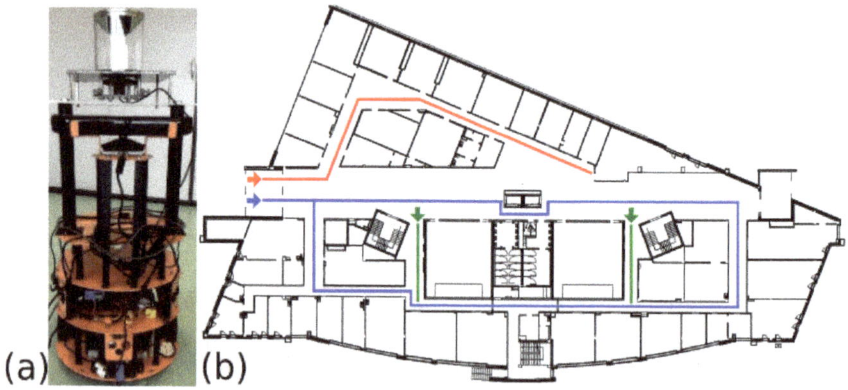

Figure 3. Labbot mobile robot with the integrated sensor with a catadioptric camera (**a**); robot paths during image collection—different colours indicate different paths (**b**).

The considered dataset contains 606 images (Figure 4a,b), which were recorded on three floors of the Poznań University of Technology Mechatronics Centre building (Figure 3b). All images were subjected to a masking process (Figure 4c) to remove areas that did not contain useful information. Using the localization system described in Section 3, embeddings of 2048 × 1 in size were calculated for each image and registered in a database of 2048 × n in size, which is a global map based on n reference images (n = 484 in the experiment). The robot's main localization task uses the integrated sensor's Jetson platform in real time. The configuration of the localization system in this experiment was the following:

- Raw catadioptric images were used (cf. Figure 4) without converting them to panoramic images.
- The neural network used to produce the embeddings was EfficientNet, which was selected upon literature-based analysis.

The EfficientNet architecture has gained prominence as an effective solution for image processing on edge devices due to its remarkable balance between accuracy and efficiency. By leveraging techniques like compound scaling, which uniformly scales the network width, depth, and resolution, EfficientNet optimizes the model's architecture to maximize accuracy while minimizing the number of parameters and computations. This enables real-time inference and efficient utilization of resources on edge devices, ensuring faster and more responsive image processing capabilities even with limited computing power [56]. Moreover, in the considered application, the input size of the available pretrained EfficientNet B5 models matches the resolution of our target images.

The used EfficientNet B5 has 577 layers and the input image size is (456,456,3). This network has high accuracy with a relatively small number of model parameters, which positively affects the processing speed of the embedded system. The network was fine-tuned before use because EfficientNet B5 was pretrained on images from the ImageNet dataset. This process was implemented using a dataset of about 10,000 augmented omnidirectional images produced from the previously collected database of 606 original images. Only standard augmentation methods available in the TensorFlow environment were applied to the images.

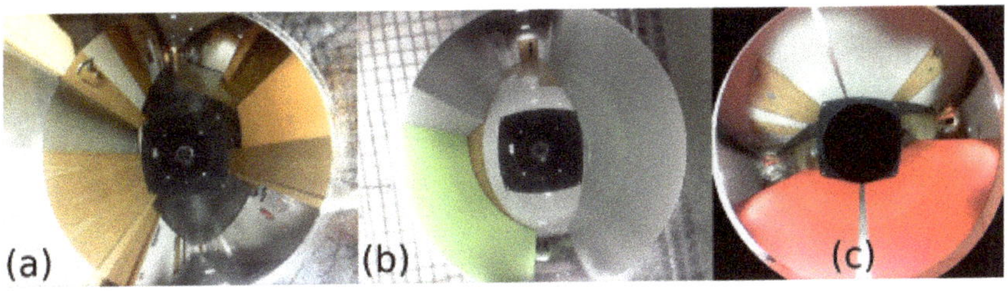

Figure 4. Omnidirectional images of different locations (**a**,**b**) in the Mechatronics Centre and an example image after masking (**c**).

A practical problem in the scenario considered in Experiment 1 was the high similarity of the indoor environment in the Mechatronics Centre building. Images were obtained approximately every 0.5 m along the robot's path, and adjacent images in the database are very similar and often indistinguishable, even by a human. Therefore, the entire dataset was manually divided into 17 different sections, each describing a topologically different location (represented by different colours in Figure 3b). Due to this organization of the dataset, no ground truth positions are provided for particular images, and we can assess the localization results only in terms of the image retrieval accuracy for particular sections. The localization process is then performed only for these 17 locations, with each location represented by 30 to 40 acquired images that partially overlap. In the training process, each section was divided into training sequences (60%), validation sequences (20%), and test sequences (20%).

4.2. Experiment 2: Stand-Alone Catadioptric Camera

The good results obtained in the preliminary experiment with the mobile robot motivated us to extend this research with a catadioptric camera of a different mechanical design and better parameters, as the relatively small horizontal field of view and often blurred images were the main drawbacks in the previous experiment.

The field of view of a catadioptric camera depends on the shape and size of the mirror being used [57]. A catadioptric sensor captures a wider field of view by using lenses and mirrors that need to be arranged carefully. Designing the mirrors is crucial to ensure a single effective viewpoint, which is necessary for generating pure perspective images from the sensed images [58]. In the new experiment, the integrated sensor was replaced by a catadioptric vision sensor consisting of a professional Basler acA2440-35uc camera with a Kowa 4.4–11 mm lens [59] and a hyperbolic mirror, whose field of view is much larger than the mirror used in the previous experiment. A hyperbolic mirror allows us to obtain the single effective viewpoint of the camera–mirror system using typical camera lenses [60], while the mirror we use in this design is larger than the previous one, and is attached at a larger distance from the camera. Both these factors contribute to a much larger horizontal field of view.

Images with a resolution of 1080×1440 were taken for two floors of the same Mechatronics Centre building: the first floor was divided into 144 places (Figure 5a) and the 3rd floor into 106 places (Figure 5b). Ground truth positions of the acquired images were obtained by measuring the position of the sensor manually with tape with respect to the known floor plan of the Mechatronics Centre building. Due to the augmentation process (Figure 6), the collection of images for training purposes increased to about thirty thousand images (details are given in Table 1).

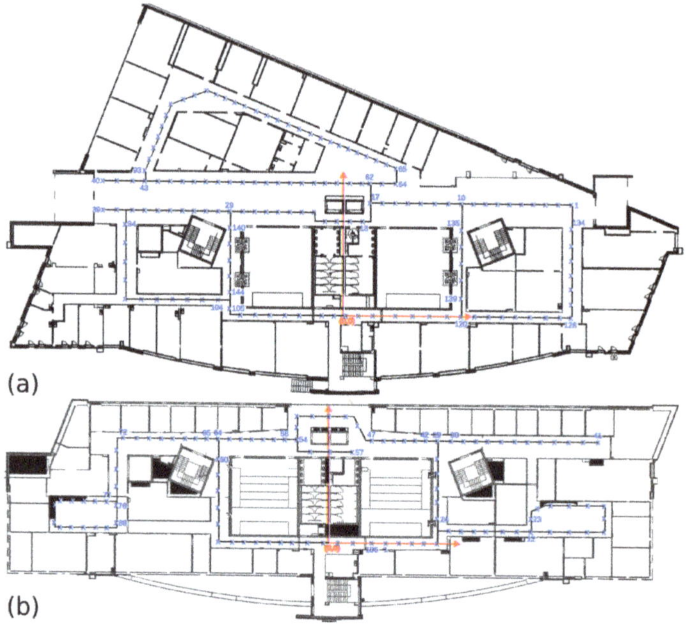

Figure 5. Blueprint of the first floor (**a**) and third floor (**b**) of the Mechatronics Centre building, with marked places (blue crosses) where images were taken.

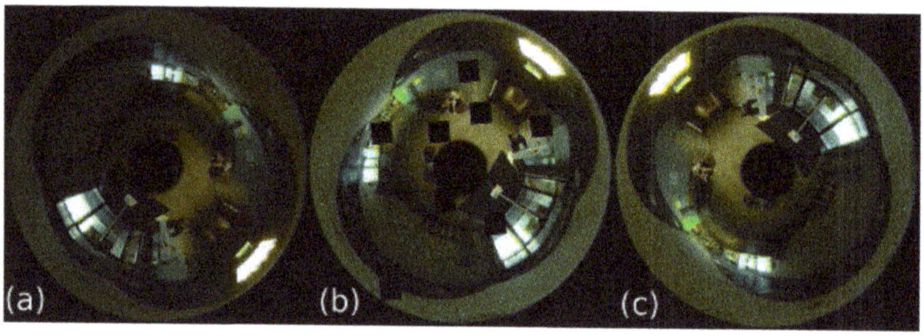

Figure 6. Example of omnidirectional image augmentation: (**a**)—original picture; (**b**,**c**)—augmented images.

In this experiment, the collected set of images was used to compare the quantitative results of place recognition for panoramic and omnidirectional images for three different CNN architectures: EfficientNet B7 [61], EfficientNet V2L [62], and MobileNetV2 [48]. Moreover, to investigate different strategies for creating the reference database (global map), the dataset was tested in three different configurations:

- Configuration A—the entire dataset was divided into a training set (60%), a validation set (20%), and a test set (20%) for each place. The validation set was then used as the reference database of embeddings.
- Configuration B—the entire dataset was divided into a training set (60%), a validation set (20%), and a test set (20%) in such a way that the locations next to the places represented in the test set were always represented in the map of embeddings. The global map of embeddings was created from a combination of the training and the validation set, but the places from the test set, used then as queries, were not directly represented in the map.
- Configuration C—all images of the places located on the first floor were divided into a training set (80%) and a validation set (20%). The set of images recorded on the third

floor was used to test the proposed solution. The 106 places for which images were recorded on the third floor were divided into the database of embeddings (80%) and a test set used as queries (20%), in such a way that the locations next to the places included in the test set were represented in the map of embeddings.

For each network configuration and dataset, the training process was conducted as in Experiment 1, with the pretrained backbone network, and by fine-training the last layers of this network on the target training dataset constructed according to the concept defined above for the given configuration.

Table 1. Number of images in training and validation datasets for Configurations A, B, and C. Numbers in brackets denote the number of images after augmentation.

	Configuration A		Configuration B		Configuration C	
	Training Dataset	**Validation Dataset**	**Training Dataset**	**Validation Dataset**	**Training Dataset**	**Validation Dataset**
omnidirectional	994 (25,844)	250 (6500)	959 (24,934)	241 (6266)	753 (19,578)	288 (7488)
panoramic	2982 (77,532)	750 (19,500)	2611 (67,886)	653 (16,978)	2259 (58,734)	864 (22,464)

As a follow-up of this experiment, we tested with the same dataset the NetVLAD architecture for comparison with our approach. It was shown in [35] that the NetVLAD architecture achieves the best-placed recognition results with the AlexNet and VGG-16 used as backbone networks. Hence, in order to compare our approach to place recognition, which is relatively simple, to the state-of-the-art NetVLAD architecture, we used a Python language implementation of NetVLAD [63] with the VGG-16 backbone. The NetVLAD model was subject to the same training process as in the case of our system, with the training sets defined in Configurations A, B, and C.

4.3. Experiment 3: COLD Datasets

An important related work to our research is the article by Cabrera Mora et al. [43], which presents several different configurations of the AlexNet network producing embeddings used for appearance-based localization with omnidirectional images without panoramic conversion. The task of the trained neural network is to perform rough localization (room identification) and then metric localization for the identified room by searching for the place closest to the query embedding. The experiments presented in [43] used images available in the Freiburg dataset, which is part of the publicly available COsy Localization Database [55]). This inspired us to replicate some of the experiments from [43] using our approach to localization. Using the same dataset and experiment design gives a chance for a fair comparison of quantitative results, which is usually not available in the not-so-common research on localization with omnidirectional images.

Moreover, we consider COLD Freiburg an interesting dataset on its own, as it contains omnidirectional images captured by a robot that followed a number of different paths in a building at the University of Freiburg. The robot visited various rooms, such as the kitchen, corridors, printer areas, bathroom, and offices (Figure 7). These rooms have wide windows and glass walls, making visual localization a particularly challenging task. The collection of images was collected under real conditions, e.g., changes in furniture, people being on the move, changes in lighting conditions (cloudy days, sunny days and nights), etc. Moreover, the images were captured while the robot was moving; therefore, they may contain blurring effects or other dynamic changes. What is important is that accurate ground truth positions of the captured images are provided in this dataset thanks to the laser scanner localization of the robot. The ground truth positions were used exclusively to measure the metric localization errors.

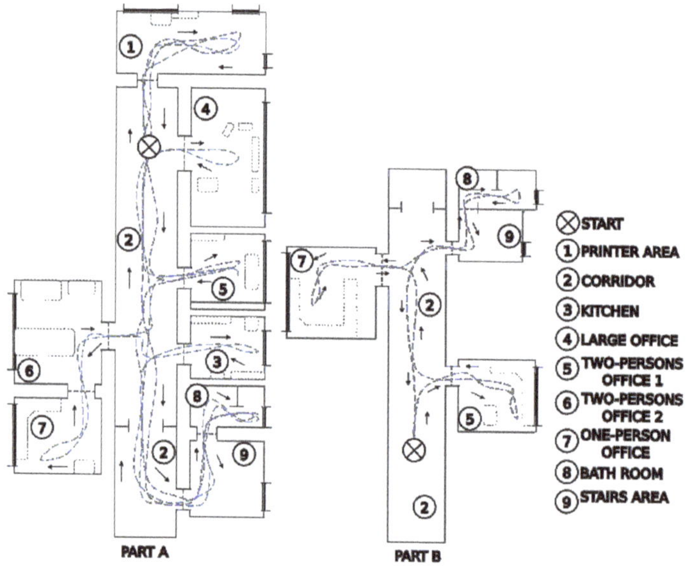

Figure 7. Maps of the two parts of the laboratory in Freiburg with approximate paths followed by the robot during data acquisition (map and trajectories data adopted from the COLD dataset web page https://www.cas.kth.se/COLD/cold-freiburg.html).

In order to evaluate the influence of the changing lighting conditions on the localization task, it was proposed in [43] to use as training data only images recorded on cloudy days, whose acquisition locations are about 20 cm apart. On the other hand, in order to assess the robustness of the location to changes in illumination, images captured on sunny and cloudy days and at night were used for testing. The COLD Freiburg dataset contains images captured in nine different rooms: a kitchen, a bathroom, a printer area, a stairwell, a long corridor, and four offices (Figure 8).

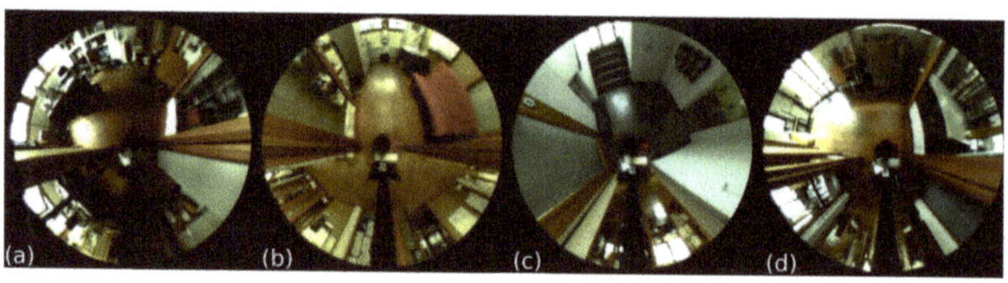

Figure 8. Example images from the COLD Freiburg dataset: (**a**) one-person office (1PO-A); (**b**) kitchen (KT-A); (**c**) stairs area (ST-A); (**d**) printer area (PA-A).

In order to evaluate the appearance-based localization system proposed in this paper, a direct comparison was made with the solution presented in [43]. To facilitate a fair comparison, we made an attempt to replicate the sets of images used in the experiments described in [43]. However, starting from the same images of the Freiburg dataset, we used only our own processing pipeline; in particular, each training set was augmented by darkening random portions of the images, rotating them, and changing the illumination. In our case, the training set was also the global map of embeddings.

Training dataset number one, which is an exact replication of the dataset from the work of [43], was obtained from a set of images taken during a cloudy day, and it was downsampled to obtain a set of images describing locations at an average distance of 20 cm between the acquisition points of successive images along the robot's path. Detailed information on the number of images contained in the training set depending on the type of lighting conditions and room is provided in Tables 2 and 3. Verification of the correctness of the obtained model for embedding generation was carried out for three test sets of query images: the first set consists of images captured on cloudy days but not included in the training set (2595 images); the second test set contains all images captured on sunny days (2807 images); and the third test set consists of all images captured at night (2876 images).

Table 2. The number of images of each room depending on the weather for three training sets, where n.i. is the number of images and % is the percentage of images of a particular room depending on the weather conditions. Dataset 1—a set of images recorded only on cloudy days. Dataset 2—set 1 extended by the missing acquisition points found in the sets for sunny days and nights. Dataset 3—a set of images showing acquisition points located every 20 cm for images recorded in all types of weather conditions.

	Training Dataset 1 (575 Images)		**Training Dataset 2 (820 Images)**						**Training Dataset 3 (1801 Images)**					
	Cloudy		**Cloudy**		**Sunny**		**Night**		**Cloudy**		**Sunny**		**Night**	
	n.i.	**%**	**n.i.**	**%**	**n.i.**	**%**	**n.i.**	**%**	**n.i.**	**%**	**n.i.**	**%**	**n.i.**	**%**
Room	**575**	**100**	**576**	**70.2**	**139**	**17.0**	**105**	**12.8**	**573**	**31.8**	**651**	**36.2**	**577**	**32.0**
1PO-A	45	100	47	69.1	15	22.0	6	8.8	46	31.3	54	36.7	47	33.0
2PO1-A	52	100	50	79.4	8	12.7	5	8.0	48	36.9	47	36.3	35	26.9
2PO2-A	33	100	30	58.8	8	15.7	13	25.5	34	30.4	40	35.7	38	33.9
CR-A	248	100	249	76.9	43	13.3	32	9.9	247	33.2	267	35.9	229	30.8
KT-A	43	100	41	42.3	31	32.0	25	25.8	40	19.9	79	39.3	82	40.8
LO-A	32	100	31	62.0	12	24.0	7	14.0	34	33.7	35	34.7	32	31.7
PA-A	58	100	58	82.9	8	11.4	4	5.7	58	37.2	55	35.3	43	27.7
ST-A	31	100	33	76.7	5	11.6	5	11.6	31	31.3	36	36.4	32	32.3
TL-A	33	100	37	68.5	9	16.7	8	14.8	35	31.3	38	33.9	39	34.9

Table 3. The number of images of each room depending on its type in a given training set, where n.i. is the number of images and % is the percentage of images of a given room depending on weather conditions. Training dataset 1—a set of images recorded only on cloudy days. Training dataset 2—dataset 1 extended by the missing acquisition points found in the datasets for sunny days and nights. Training dataset 3—a set of images showing acquisition points located every 20 cm for images recorded in all types of weather conditions.

	Training Dataset 1 (575 Images)		**Training Dataset 2 (820 Images)**		**Training Dataset 3 (1801 Images)**	
Room.	**n.i.**	**%**	**n.i.**	**%**	**n.i.**	**%**
1PO-A	45	7.83	68	8.29	147	8.16
2PO1-A	52	9.04	63	7.68	130	7.22
2PO2-A	33	5.74	51	6.21	112	6.22
CR-A	248	43.13	324	39.51	743	41.25
KT-A	43	7.48	97	11.83	201	11.16
LO-A	32	5.57	50	6.1	101	5.61
PA-A	58	10.09	70	8.54	156	8.66
ST-A	31	5.39	43	5.24	99	5.50
TL-A	33	5.74	54	6.59	112	6.22

5. Results and Discussion

This section presents and discusses the results of the three experiments described in this paper. Quantitative results in terms of place recognition (i.e., image retrieval) accuracy are presented for all experiments. For experiments no. 2 and no. 3, we also present quantitative results in terms of the metric localization accuracy, as the datasets used in these experiments provide ground truth for positions of the place images in a global reference system. Moreover, we discuss qualitative localization results, pointing out the most common sources of localization errors and providing recommendations for training strategies of the deep neural networks that make the resulting models robust to changes in the environment.

5.1. Experiment 1

As defined in Section 4, in Experiment 1, appearance-based localization was conducted for 17 sections, each of them containing several image acquisition locations, and being a description of a larger corridor space. The best network training results were obtained for the unfrozen last 50 layers of the backbone CNN, a learning rate of 1×10^{-4}, and a batch size of 16, with a learning error of 0.1605, learning accuracy of 0.9596, validation error of 0.1183, and validation accuracy of 0.9796 (Figure 9).

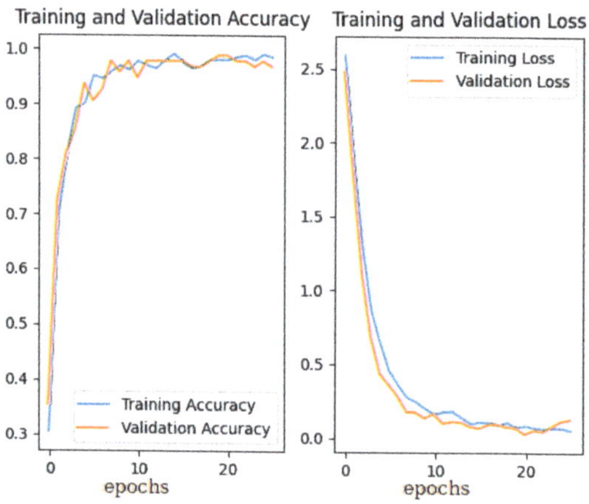

Figure 9. Model training results in Experiment 1.

On the test dataset containing 122 query images, the average accuracy of place recognition was 98%, with very few misclassified queries, as shown by the confusion matrix in Figure 10. The average processing time of a single query image was 480 ms, with a standard deviation of 83ms and a maximum time of 1313 ms, allowing for real-time localization. A qualitative example of place recognition is given in Figure 11. Visual inspection of the results and the confusion matrix suggest that the most common section mismatch is when the same place is at the start of a new section and the end of a previous section. However, errors also are caused by blurred images and bright spots of sunlight or artificial light in the images.

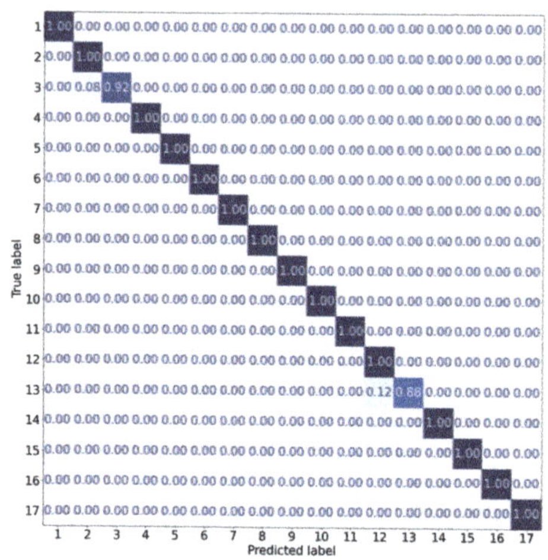

Figure 10. Confusion matrix for 17 sections.

This experiment allowed us to conclude that the proposed approach to appearance-based localization with embeddings produced by a lightweight CNN suits the target application in terms of both image retrieval accuracy and real-time performance. However, the used sensor, having rather a small field of view and mechanical structure prone to decalibration and defocusing (causing blurred images), did not allow us to extend these investigations to a larger dataset with ground truth positions of images.

Figure 11. Results of sample section predictions. The image in the first column is a query; the other columns are the four closest neighbours. In square brackets, there is the section number (i.e. [12], [02]), and next to it, the L2 distances between the query and the presented image are given. An example of (**a**) correct place recognition and (**b**) mismatched sections having slightly overlapping ranges.

5.2. Experiment 2

The conclusions drawn from Experiment 1 were taken into account while designing the next experiment involving the use of a different catadioptric sensor and a more extended and diversified dataset of indoor images with ground truth positions.

In Experiment 2, the convolutional neural networks EfficientNet B7, EfficientNet V2L, and MobileNet V2 were compared for Configurations A, B, and C. The quantitative results presented in the graphs show the percentage of Cartesian locations found in the given distance intervals expressed in meters, and the average Euclidean distance measurement error: for Configuration A—Figure 12; Configuration B—Figure 13; and Configuration C—Figure 14. Moreover, Table 4 shows the average time of processing a single query image with the proposed solution based on embeddings and Faiss KNN search.

Table 4. Mean Euclidean distance error ($\overline{\overline{b_L}}$) and mean time ($\bar{\bar{t}}$) of location determination on the Jetson TX2 computing platform for original (omnidirectional) and panoramic images.

		Experiment 2								
Neural Network	Image Type	Configuration A			Configuration B			Configuration C		
		$\overline{\overline{b_L}}$ [m]	$\bar{\bar{t}}$ [s]	$\bar{\bar{t}}_{tr}$ [h]	$\overline{\overline{b_L}}$ [m]	$\bar{\bar{t}}$ [s]	t_{tr} [h]	$\overline{\overline{b_L}}$ [m]	$\bar{\bar{t}}$ [s]	t_{tr} [h]
EfficientNet B7	omni	0.00	0.52	2.15	3.06	0.48	3.25	4.43	0.47	2.16
EfficientNet B7	panoramic	0.03	0.56	37.21	3.21	0.49	16.24	3.92	0.50	11.30
EfficientNet V2L	omni	0.00	0.35	1.98	2.34	0.35	3.84	4.94	0.34	2.07
EfficientNet V2L	panoramic	0.00	0.39	14.54	3.11	0.37	15.46	3.60	0.36	12.14
MobileNet V2	omni	0.02	0.08	2.24	3.86	0.07	3.15	5.01	0.07	1.55
MobileNet V2	panoramic	0.36	0.11	16.32	4.33	0.11	15.56	6.87	0.11	11.53

A mismatch of the neighbouring places occurs only when the images overlap significantly (note that the catadioptric camera used in Experiment 2 has a much larger field of view that the previously used one) and are very similar to each other due to the self-similar nature of the environment. No significant difference was noticed between the results obtained for omnidirectional and panoramic images, which indicates that for appearance-based localization with our approach, it is unnecessary to convert the omnidirectional images to panoramic images. Hence, we can avoid the time-consuming conversion and rectification procedure [8], without compromising the results.

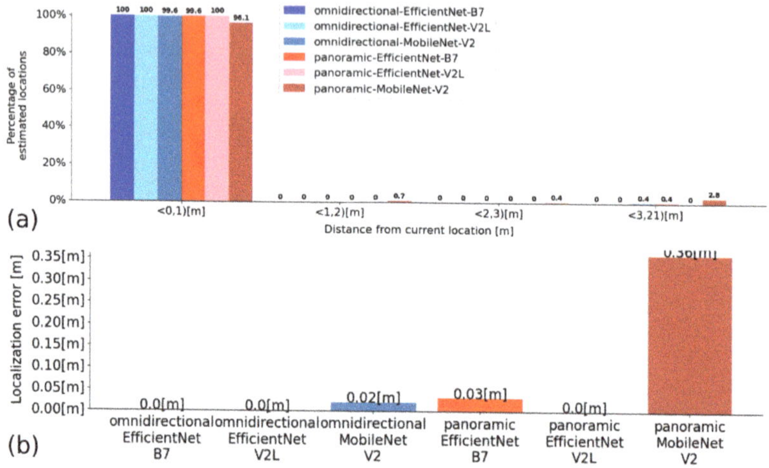

Figure 12. Quantitative results for Configuration A: (**a**)—a percentage of matches that are within a range of distance from the actual distance (the units on the x-axis are the ranges of distances); (**b**)—average distance measurement error.

As a follow-up of Experiment 2, a comparison of localization results was performed between the NetVLAD approach with VGG-16 and VLAD layer, and our approach with two variants of the EfficientNet backbone. This comparison was performed for all three configurations of the reference map (A, B, and C) and both the omnidirectional and converted panoramic images.

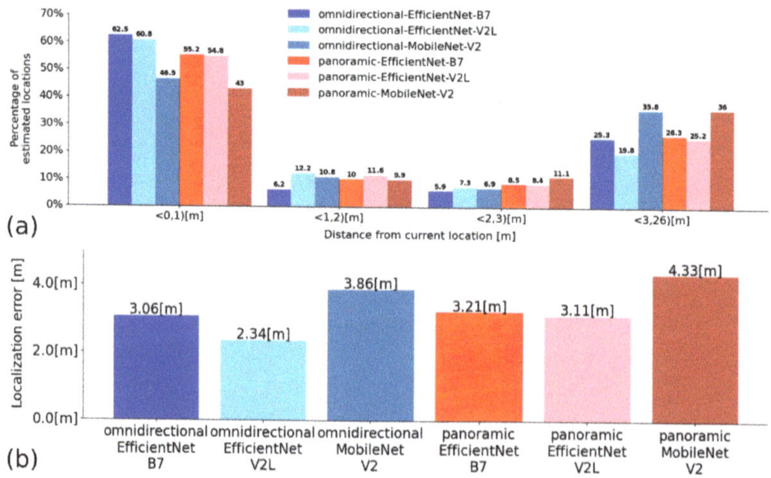

Figure 13. Quantitative results for Configuration B. (**a**)—the percentage of matches that are within a range of distance from the actual distance (the units on the x-axis are the ranges of distances); (**b**)—the average distance measurement error.

Quantitative results of the average Euclidean distance measurement error for all configurations considered in this test are shown in Table 5. As can be seen in this table, for the panoramic images, the proposed solution has a smaller average error for distance measurement than NetVLAD. On the other hand, for original omnidirectional images, the proposed solution with EfficientNet and embedding has the same or larger average error for distance measurement than NetVLAD with VGG-16.

From Experiment 2, we conclude that the CNN architecture performing best in our system is EfficientNet V2L, a model from a recently introduced family of convolutional networks that achieve faster training and better parameter efficiency than older network models [62]. This model, being up to 6.8 times smaller than state-of-the-art models, suits our embedded computing platform well. Moreover, our approach performs as a pair with the much more complicated and much bigger NetVLAD architecture. On the other hand, Experiment 2 shows that NetVLAD can handle raw omnidirectional images without converting them to panoramic images if it is trained on a representative dataset. As to the strategy of defining the reference map and the training dataset, the results of Experiment 2 show that it is possible to find a correct neighbouring place, even if the very exact image of the queried place is not included in the reference map. However, these results also show that the generalization ability of the investigated deep learning solutions, including NetVLAD, is somewhat limited if the query images come from a different environment than the training set. This is suggested by the worse results in Configuration C, no matter if omnidirectional or panoramic images are being used. Therefore, the generalization ability of the proposed architecture should be further investigated.

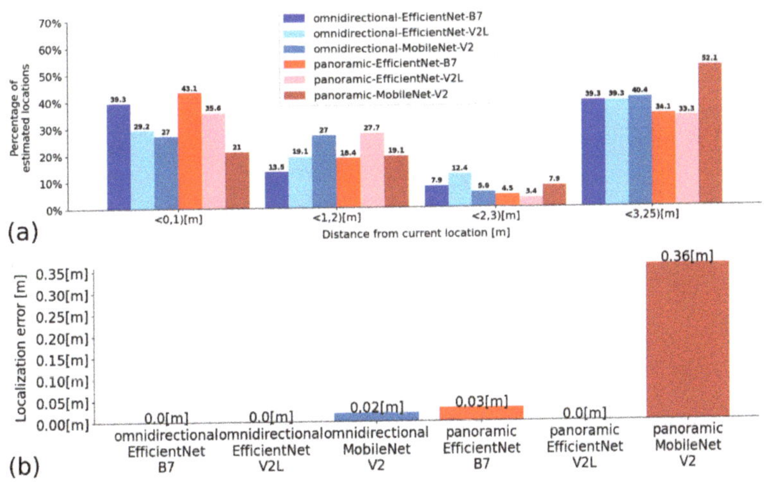

Figure 14. Quantitative results for Configuration C: (**a**)—the percentage of matches that are within a range of distance from the actual distance (the units on the x-axis are the ranges of distances); (**b**)—the average distance measurement error.

Table 5. Comparison of mean Euclidean distance error for the network architecture presented in this paper (EfficientNet B7/EfficientNet V2L + embeddings) and NetVLAD network for omnidirectional and panoramic images.

	Experiment 2					
	Configuration A		Configuration B		Configuration C	
	Omni	Panoramic	Omni	Panoramic	Omni	Panoramic
Neural Network	$\overline{\overline{b_L}}$ [m]	$\overline{\overline{b_L}}$ [m]	$\overline{\overline{b_L}}$ [m]	$\overline{\overline{b_L}}$ [m]	$\overline{\overline{b_L}}$ [m]	$\overline{\overline{b_L}}$ [m]
EfficientNet B7 + embeddings	0.00	0.03	3.06	3.21	4.43	3.92
EfficientNet V2L + embeddings	0.00	0.00	2.34	3.11	4.94	3.60
NetVLAD (VGG16 + VLAD)	0.00	0.10	2.27	3.77	2.24	4.60

5.3. Experiment 3

Conclusions about the limited generalization ability of the proposed localization system drawn from Experiment 2 were one of the main motivations behind the concept of Experiment 3, which applies the publicly available COLD Freiburg dataset and compares side-by-side to the results obtained on the same dataset and published recently in [43].

In [43], the best results were obtained for the retrained AlexNet, with 97.11% of correct room identifications obtained for images captured during cloudy daytime, 93.48% for sunny days, and 96.77% for nighttime.

Following the methodology of replicating the selected experiments from [43] with respect to the used images, we achieved the following results for appearance-based localization using embeddings with EfficientNet V2L: 98.03% for cloudy days, 97.01% for sunny days, and 97.77% for nights. The full quantitative results are presented in Figure 15. Thus, better accuracy of room recognition was obtained for cloudy days by 0.92%, by 3.53% for sunny days, and by 1% for images recorded at night. The average distance measurement error for the room was also analysed, and the results obtained are presented in Figure 15b. The proposed solution achieved a smaller average distance measurement error than in [43] for cloudy days by 0.05 m, and by 0.24 m for sunny days; only for images taken at night was the average error increased by 0.04 m.

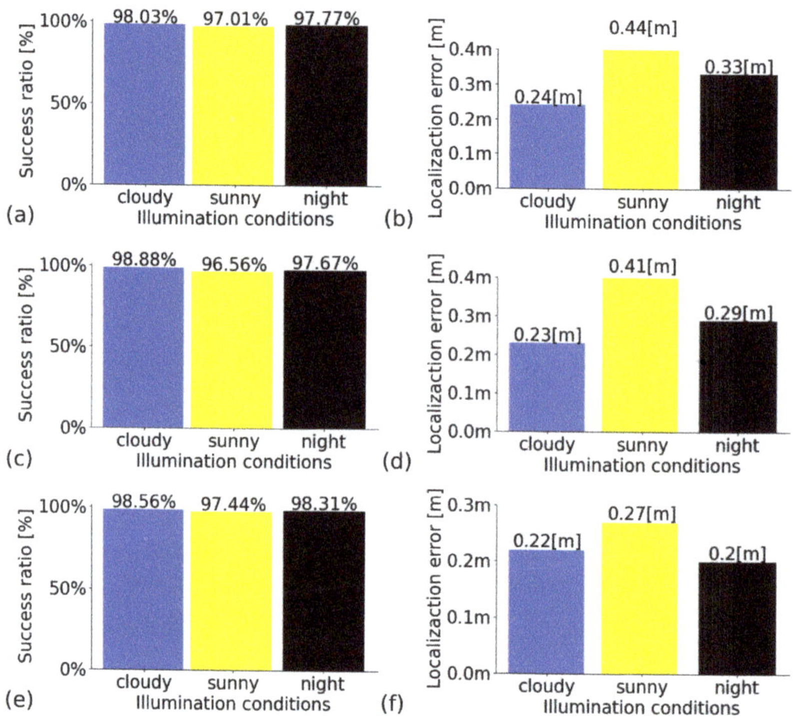

Figure 15. Success ratio for EfficientNetV2L and the set of embeddings acquired from the training set for the room search task for the COLD Freiburg dataset. The result obtained under cloudy (blue), night (black), and sunny (yellow) conditions for the model learned on the training set, namely, a set of images on cloudy days (**a**), a set of images on cloudy days extended by missing acquisition locations found in images for sunny days and night (**b**), and a balanced set of images obtained on cloudy and sunny days and at night (**c**). Average location error in meters for a set of images on cloudy days (**d**), a set of images on cloudy days extended by missing acquisition locations found in images for sunny days and at night (**e**), and a balanced set of images obtained on cloudy and sunny days and at night (**f**).

During the analysis of the results, it was noticed that the images recorded on sunny days and nights have different acquisition spots (positions) than those acquired for cloudy days, and some of them are farther than 20 cm from the acquisition spots for cloudy days, e.g., for the room labelled CR-A, the farthest acquisition spot on sunny days was 0.5 m away from the farthest acquisition spot on a cloudy day. For this reason, an extension of the training set no. 1 was made, only with images taken at the missing acquisition spots for sunny days and for nighttime (Tables 2 and 3). Note that the images added to the learning set were removed from the test sets, and never used as queries. The results obtained with this amendment of the training set are shown in Figure 15c,d. However, neither in the case of room identification accuracy nor for position determination errors did the results improve significantly.

Based on the data from Table 2, it was noticed that the training dataset number 2 is not balanced in terms of images of a given room depending on lighting conditions. In deep learning, several techniques are commonly used to handle imbalanced datasets of images. One approach is oversampling, where the minority class samples are replicated to match the majority class. Another technique is undersampling, where random samples from the majority class are removed to balance the dataset. Additionally, there are methods like synthetic data generation, cost-sensitive learning, and ensemble techniques [64]. However, these methods have certain drawbacks. Oversampling can lead to overfitting and a loss of generalization ability. Undersampling can discard valuable information and result in underrepresented classes being ignored. Synthetic data generation may introduce unrealistic patterns. Cost-sensitive learning requires careful tuning of class weights. There is no technique that universally addresses all imbalanced dataset challenges. For this reason, a third training set was created (no. 3), which contains images representing acquisition locations about 20 cm apart for each room at night and on sunny and cloudy days (Tables 2 and 3). Finally, for this illumination-balanced training set, the best results were obtained for room identification: 98.56% for cloudy days, 97.44% for sunny days, and 98.31% for nights (Figure 15e). Also, the smallest average distance measurement error was obtained for training with the set no. 3: 0.22 m for cloudy days, 0.27 m for sunny days, and 0.2 m for nights (Figure 15f). Based on the data from Table 3, it can be seen that the percentages of the images of each room in each training set are very close to each other, but only by balancing the set by the lighting conditions (i.e., sunny, cloudy, night) were satisfactory results obtained.

In order to explore the generalizability of the proposed system to other environments, we also demonstrate localization results for the Saarbrücken sequence from the COLD dataset. This is a sequence of images collected in a different location within a building having different characteristics than the one used for the Freiburg sequence [20]. Both Freiburg and Saarbrücken sequences from the COLD dataset were used in the research on mobile robot's localization presented in [32], although the inaccurate description of the dataset used in this paper does not allow us to replicate the Saarbrücken sequence experiment for direct comparison, as it was accomplished for the Freiburg sequence used in [43]. Therefore, considering the analysis of the influence of an imbalanced training dataset on the localization accuracy of the system, it was decided to use only part B of the Saarbrücken sequence, since it is the only one with images for all weather and lighting conditions. The results presented in Figure 16 confirm that our approach generalizes well to different indoor environments, achieving a similar success rate of image retrieval and a similar average localization accuracy as the Freiburg sequence.

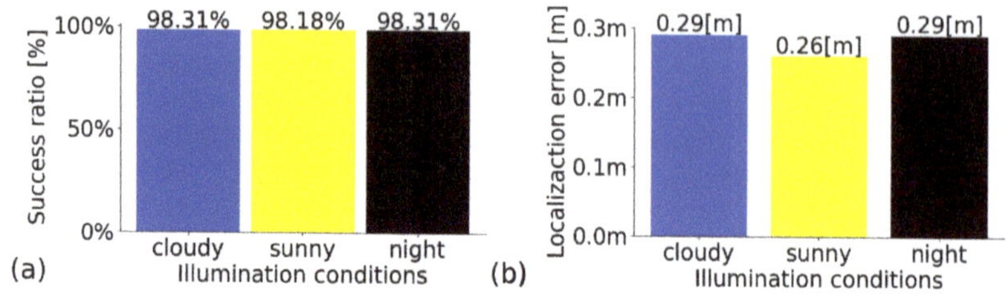

Figure 16. Success ratio for EfficientNetV2L and the set of embeddings acquired from the training set for the room search task for COLD Saarbrücken dataset for part B. Results obtained under cloudy (blue), night (black), and sunny (yellow) conditions for the model learned on the training set are a balanced set of images obtained on cloudy and sunny days and night (**a**). Average location error in meters for a balanced set of images obtained on cloudy and sunny days and night (**b**).

The hierarchical localization of the mobile robot in [43] is divided into two subtasks: the first one is room recognition, and the second one is accurate localization within the known room. Accurate localization involves estimating the position in which the test image was captured using a nearest neighbour search method in the space of global descriptors obtained from the convolutional network, which is similar to our approach. In the paper, several experiments with the COLD dataset were performed based on different configurations of the networks and training sets used in the room classification task. Finally, the authors of [43] selected the CNN network trained in experiment number 2 and the fully connected layer numbered 6 in their network as the one producing the global descriptors from images, because this configuration showed the greatest robustness to changes in lighting conditions in the preliminary experiments. This configuration is referred to as CNN2 + FC6. Another configuration considered in [43] is the CNN network trained in experiment 8 and the output of the sixth layer (CNN8 + FC6).

Table 6 presents the quantitative results and computation times obtained for our appearance-based localization. The average processing time per image was 0.32 seconds (achieving 3.13 FPS), making appearance-based localization in real-time in the context of the typical motion speed of our service robot. For comparison purposes, this table also contains results for the two selected configurations of the system from [43], and results of other hierarchical localization methods (not learning-based) investigated in [32] and used for comparison in [43]. Note that the numerical results from [32,43] are taken from the respective papers, as we did not attempt to reimplement these systems, replicating only the training and test sets for Experiment 3.

Table 6. Comparison of results for different approaches to place recognition (appearance-based localization) using images from a catadioptric camera. Best results shown in bold.

Global Descriptor	$\overline{\overline{b_L}}$ **[m]** **Cloudy**	$\overline{\overline{b_L}}$ **[m]** **Sunny**	$\overline{\overline{b_L}}$ **[m]** **Night**
EfficientNet V2L (training dataset 3)	0.22 ($\bar{\bar{t}} = 0.32$ s)	**0.27** ($\bar{\bar{t}} = 0.31$ s)	**0.20** ($\bar{\bar{t}} = 0.31$ s)
EfficientNet V2L (training dataset 1)	0.24 ($\bar{\bar{t}} = 0.32$ s)	0.44 ($\bar{\bar{t}} = 0.32$ s)	0.33 ($\bar{\bar{t}} = 0.32$ s)
CNN2 + FC6 [43]	0.29	0.69	0.29
CNN8 + FC6 [43]	0.25	0.93	0.24
HOG [32]	0.31	1.57	0.95
GIST [32]	**0.08**	1.23	1.31

The appearance-based localization methods studied in [32] include the global HOG and Gist descriptors. HOG is a feature decoder often used in image processing for object detection. The Gist descriptor, on the other hand, is used to extract global features of the environment by combining visual and semantic information through a set of perceptual dimensions that represent the dominant spatial structure of the scene. Both HOG and

GIST are descriptors that do not use machine learning algorithms, providing an interesting reference point for the method proposed in the article.

The results shown in Figure 15 and in Table 6 document that the approach to appearance-based localization proposed in this paper, despite being simple and lightweight, outperforms some recently published results in this area. Considering the fact that our solution turned out to be similar in performance to the state-of-the-art NetVLAD deep learning architecture, which requires many more computing resources, we conclude that our research reached its goal, demonstrating a versatile and accurate deep learning architecture that is suitable for the low-cost Nvidia Jetson TX2 computing platform. However, the main scientific outcome of Experiment 3 seems to be demonstrating how important it is to prepare a balanced dataset for training, particularly in the context of generalization over various image acquisition characteristics.

6. Conclusions

The results of the tests of the place recognition software for catadioptric cameras and the edge computing platform allow us to conclude that the proposed neural network architecture and parallel processing make it possible to obtain a real-time localization system that works with raw catadioptric images, despite their distorted nature.

The extensive study of the algorithm of appearance-based localization and comparison of results with similar solutions known from the literature demonstrate that the proposed approach makes it possible to obtain highly descriptive embeddings of the observed locations, and consequently, efficient appearance-based localization.

The most important conclusions, summarizing the remarks discussed in Section 5, concern the best performance of the EfficientNet V2L CNN backbone for generating the embeddings and the pivotal importance of preparing a well-balanced training set for this network, even if transfer learning with pretraining on a large dataset of general purpose images is used. A practical conclusion is that the not-so-recent and low-cost Nvidia Jetson TX2 embedded computer is enough to run a carefully engineered deep learning system for appearance-based localization. This opens interesting opportunities for developing affordable service and social indoor mobile robots utilizing a catadioptric camera as the main localization sensor.

However, a limitation of the proposed appearance-only approach to global localization is the limited accuracy of the obtained metric position of the robot. This accuracy depends on the density of the global map, because the obtained position of the robot is defined by the known location of the most similar image. If the images were collected close to each other, then the position of the robot can be determined more accurately, but if the distances between the points where the images were captured are large, the accuracy is decreased. This limitation will be addressed in our further research by implementing a neural network that will regress the position of the robot with respect to the reference image retrieved from the map. Further research on this system will also concern the implementation of triplet loss with hard negative mining, as this training scheme turned out to be very effective in a number of localization systems. This training strategy should allow the network to develop more specific features, thus making the localization system more effective in highly repetitive indoor environments.

Author Contributions: Conceptualization, M.R.; methodology, P.S.; software, M.R.; validation, M.R.; formal analysis, M.R.; investigation, M.R.; resources, P.S.; data curation, M.R.; writing—original draft preparation, M.R. and P.S.; writing—review and editing, M.R. and P.S.; visualization, M.R.; supervision, P.S.; project administration, P.S.; funding acquisition, P.S. All authors have read and agreed to the published version of the manuscript.

Funding: This research was funded by Poznań University of Technology, internal grant number 0214/SBAD/0242.

Institutional Review Board Statement: Not applicable.

Informed Consent Statement: Not applicable.

Data Availability Statement: Open source code and our datasets of images are available on GitHub: https://github.com/mrostkowska/real-time-indoor-localization-catadioptric-vision-sensor.git, accessed on 18 June 2023. The used COLD dataset is available on this project's web page (https://www.cas.kth.se/COLD/ accessed on 30 May 2023).

Conflicts of Interest: The authors declare no conflict of interest.

References

1. Lee, I. Service Robots: A Systematic Literature Review. *Electronics* **2021**, *10*, 2658. [CrossRef]
2. Zachiotis, G.A.; Andrikopoulos, G.; Gornez, R.; Nakamura, K.; Nikolakopoulos, G. A Survey on the Application Trends of Home Service Robotics. In Proceedings of the IEEE International Conference on Robotics and Biomimetics (ROBIO), Kuala Lumpur, Malaysia, 12–15 December 2018; pp. 1999–2006.
3. Asgharian, P.; Panchea, A.M.; Ferland, F. A Review on the Use of Mobile Service Robots in Elderly Care. *Robotics* **2022**, *11*, 127. [CrossRef]
4. Skrzypczyński, P.; Tobis, S. Eldercare Robots in the Age of AI: Are We Ready to Address the User Needs? In Proceedings of the 3rd Polish Conference on Artificial Intelligence PP-RAI'2022, Gdynia, Poland, 25–27 April 2022; pp. 116–121.
5. Huang, J.; Junginger, S.; Liu, H.; Thurow, K. Indoor Positioning Systems of Mobile Robots: A Review. *Robotics* **2023**, *12*, 47. [CrossRef]
6. Sousa, R.B.; Sobreira, H.M.; Moreira, A.P. A systematic literature review on long-term localization and mapping for mobile robots. *J. Field Robot.* **2023**, *40*, 1245–1322. [CrossRef]
7. Wietrzykowski, J.; Skrzypczyński, P. PlaneLoc: Probabilistic global localization in 3-D using local planar features. *Robot. Auton. Syst.* **2019**, *113*, 160–173. [CrossRef]
8. Rostkowska, M.; Skrzypczyński, P. Hybrid field of view vision: From biological inspirations to integrated sensor design. In Proceedings of the IEEE International Conference on Multisensor Fusion and Integration for Intelligent Systems (MFI), Baden-Baden, Germany, 19–21 September 2016; pp. 629–634.
9. Skrzypczyński, P.; Rostkowska, M.; Wasik, M. Bio-Inspired, Real-Time Passive Vision for Mobile Robots. In *Machine Vision and Navigation*; Springer International Publishing: Cham, Switzerland, 2020; pp. 33–58.
10. Lowry, S.; Sünderhauf, N.; Newman, P.; Leonard, J.J.; Cox, D.; Corke, P.; Milford, M.J. Visual Place Recognition: A Survey. *IEEE Trans. Robot.* **2016**, *32*, 1–19. [CrossRef]
11. Rostkowska, M.; Skrzypczyński, P. A Practical Application of QR-codes for Mobile Robot Localization in Home Environment. In *Human-Centric Robotics: Proceedings of CLAWAR 2017: 20th International Conference on Climbing and Walking Robots and the Support Technologies for Mobile Machines, Porto, Portugal, 11–13 September 2018*; World Scientific: Singapore, 2018; pp. 311–318.
12. Arroyo, R.; Alcantarilla, P.F.; Bergasa, L.M.; Romera, E. Towards life-long visual localization using an efficient matching of binary sequences from images. In Proceedings of the 2015 IEEE International Conference on Robotics and Automation (ICRA), Seattle, WA, USA, 26–30 May 2015; pp. 6328–6335.
13. Wang, T.; Huang, H.; Lin, J.; Hu, C.; Zeng, K.; Sun, M. Omnidirectional CNN for Visual Place Recognition and Navigation. In Proceedings of the IEEE International Conference on Robotics and Automation (ICRA), Brisbane, Australia, 21–25 May 2018; pp. 2341–2348.
14. Yokoyama, A.M.; Ferro, M.; de Paula, F.B.; Vieira, V.G.; Schulze, B. Investigating hardware and software aspects in the energy consumption of machine learning: A green AI-centric analysis. In *Concurrency and Computation: Practice and Experience*; Wiley: Hoboken, NJ, USA, 2023; p. e7825.
15. Süzen, A.A.; Duman, B.; Şen, B. Benchmark Analysis of Jetson TX2, Jetson Nano and Raspberry PI using Deep-CNN. In Proceedings of the International Congress on Human-Computer Interaction, Optimization and Robotic Applications (HORA), Ankara, Turkey, 26–27 June 2020; pp. 1–5.
16. Lemaire, T.; Berger, C.; Jung, I.K.; Lacroix, S. Vision-Based SLAM: Stereo and Monocular Approaches. *Int. J. Comput. Vis.* **2007**, *74*, 343–364. [CrossRef]
17. Macario Barros, A.; Michel, M.; Moline, Y.; Corre, G.; Carrel, F. A Comprehensive Survey of Visual SLAM Algorithms. *Robotics* **2022**, *11*, 24. [CrossRef]
18. Labbé, M.; Michaud, F. Appearance-Based Loop Closure Detection for Online Large-Scale and Long-Term Operation. *IEEE Trans. Robot.* **2013**, *29*, 734–745. [CrossRef]
19. Williams, B.; Cummins, M.; Neira, J.; Newman, P.; Reid, I.; Tardós, J. A comparison of loop closing techniques in monocular SLAM. *Robot. Auton. Syst.* **2009**, *57*, 1188–1197. [CrossRef]
20. Ullah, M.M.; Pronobis, A.; Caputo, B.; Luo, J.; Jensfelt, P.; Christensen, H.I. Towards robust place recognition for robot localization. In Proceedings of the IEEE International Conference on Robotics and Automation, Pasadena, CA, USA, 19–23 May 2008; pp. 530–537.
21. Nowicki, M.R.; Wietrzykowski, J.; Skrzypczyński, P. Real-Time Visual Place Recognition for Personal Localization on a Mobile Device. *Wirel. Pers. Commun.* **2017**, *97*, 213–244. [CrossRef]
22. Murillo, A.C.; Guerrero, J.J.; Sagues, C. SURF features for efficient robot localization with omnidirectional images. In Proceedings of the IEEE International Conference on Robotics and Automation, Roma, Italy, 10–14 April 2007; pp. 3901–3907.

23. Schmidt, A.; Kraft, M.; Fularz, M.; Domagala, Z. Comparative assessment of point feature detectors and descriptors in the context of robot navigation. *J. Autom. Mob. Robot. Intell. Syst. JAMRIS* **2013**, *7*, 11–20.
24. Sivic, J.; Zisserman, A. Video Google: A text retrieval approach to object matching in videos. In Proceedings of the Ninth IEEE International Conference on Computer Vision, Nice, France, 14–17 October 2003; Volume 2, pp. 1470–1477.
25. Cummins, M.; Newman, P. FAB-MAP: Probabilistic Localization and Mapping in the Space of Appearance. *Int. J. Robot. Res.* **2008**, *27*, 647–665. [CrossRef]
26. Cummins, M.; Newman, P. Appearance-only SLAM at large scale with FAB-MAP 2.0. *Int. J. Robot. Res.* **2010**, *30*, 1100–1123. [CrossRef]
27. Román, V.; Payá, L.; Peidró, A.; Ballesta, M.; Reinoso, O. The Role of Global Appearance of Omnidirectional Images in Relative Distance and Orientation Retrieval. *Sensors* **2021**, *21*, 3327. [CrossRef] [PubMed]
28. Menegatti, E.; Maeda, T.; Ishiguro, H. Image-based memory for robot navigation using properties of omnidirectional images. *Robot. Auton. Syst.* **2004**, *47*, 251–267. [CrossRef]
29. Payá, L.; Reinoso, O.; Jiménez, L.; Julia, M. Estimating the position and orientation of a mobile robot with respect to a trajectory using omnidirectional imaging and global appearance. *PLoS ONE* **2017**, *12*, e0175938. [CrossRef]
30. Dalal, N.; Triggs, B. Histograms of oriented gradients for human detection. In Proceedings of the IEEE Computer Society Conference on Computer Vision and Pattern Recognition (CVPR'05), San Diego, CA, USA, 20–25 June 2005; Volume 1, pp. 886–893.
31. Oliva, A.; Torralba, A. Chapter 2 Building the gist of a scene: The role of global image features in recognition. In *Progress in Brain Research: Visual Perception*; Martinez-Conde, S., Macknik, S., Martinez, L., Alonso, J.M., Tse, P., Eds.; Elsevier: Amsterdam, The Netherlands, 2006; Volume 155, pp. 23–36.
32. Cebollada, S.; Payá, L.; Mayol-Cuevas, W.; Reinoso, O. Evaluation of Clustering Methods in Compression of Topological Models and Visual Place Recognition Using Global Appearance Descriptors. *Appl. Sci.* **2019**, *9*, 377. [CrossRef]
33. Ai, H.; Cao, Z.; Zhu, J.; Bai, H.; Chen, Y.; Wang, L. Deep Learning for Omnidirectional Vision: A Survey and New Perspectives. *arXiv* **2022**, arXiv:2205.10468.
34. Li, Q.; Li, K.; You, X.; Bu, S.; Liu, Z. Place recognition based on deep feature and adaptive weighting of similarity matrix. *Neurocomputing* **2016**, *199*, 114–127. [CrossRef]
35. Arandjelović, R.; Gronat, P.; Torii, A.; Pajdla, T.; Sivic, J. NetVLAD: CNN architecture for weakly supervised place recognition. In Proceedings of the IEEE Conference on Computer Vision and Pattern Recognition, Las Vegas, NV, USA, 27–30 June 2016.
36. Zhang, J.; Cao, Y.; Wu, Q. Vector of Locally and Adaptively Aggregated Descriptors for Image Feature Representation. *Pattern Recognit.* **2021**, *116*, 107952. [CrossRef]
37. Jégou, H.; Douze, M.; Schmid, C.; Pérez, P. Aggregating local descriptors into a compact image representation. In Proceedings of the 2010 IEEE Computer Society Conference on Computer Vision and Pattern Recognition, San Francisco, CA, USA, 13–18 June 2010; pp. 3304–3311.
38. Gong, Y.; Wang, L.; Guo, R.; Lazebnik, S. *Multi-scale Orderless Pooling of Deep Convolutional Activation Features*; Springer International Publishing: Berlin/Heidelberg, Germany, 2014.
39. Cheng, R.; Wang, K.; Lin, S.; Hu, W.; Yang, K.; Huang, X.; Li, H.; Sun, D.; Bai, J. Panoramic Annular Localizer: Tackling the Variation Challenges of Outdoor Localization Using Panoramic Annular Images and Active Deep Descriptors. In Proceedings of the IEEE Intelligent Transportation Systems Conference (ITSC), Auckland, New Zealand, 27–30 October 2019; pp. 920–925.
40. Cebollada, S.; Payá, L.; Flores, M.; Roman, V.; Peidro, A.; Reinoso, O. A Deep Learning Tool to Solve Localization in Mobile Autonomous Robotics. In Proceedings of the 17th International Conference on Informatics in Control, Automation and Robotics, Online, 7–9 July 2020; pp. 232–241.
41. Masci, J.; Migliore, D.; Bronstein, M.M.; Schmidhuber, J. Descriptor Learning for Omnidirectional Image Matching. In *Registration and Recognition in Images and Videos*; Springer: Berlin/Heidelberg, Germany, 2014; pp. 49–62.
42. Ballesta, M.; Payá, L.; Cebollada, S.; Reinoso, O.; Murcia, F. A CNN Regression Approach to Mobile Robot Localization Using Omnidirectional Images. *Appl. Sci.* **2021**, *11*, 7521. [CrossRef]
43. Mora, J.C.; Cebollada, S.; Flores, M.; Reinoso, Ó.; Payá, L. Training, Optimization and Validation of a CNN for Room Retrieval and Description of Omnidirectional Images. *SN Comput. Sci.* **2022**, *3*, 271. [CrossRef]
44. Cunningham, P.; Delany, S.J. k-Nearest neighbour classifiers-A Tutorial. *ACM Comput. Surv. CSUR* **2021**, *54*, 1–25. [CrossRef]
45. Kramer, O. K-Nearest Neighbors. In *Dimensionality Reduction with Unsupervised Nearest Neighbors*; Springer: Berlin/Heidelberg, Germany, 2013; pp. 13–23.
46. Ab Wahab, M.N.; Nazir, A.; Zhen Ren, A.T.; Mohd Noor, M.H.; Akbar, M.F.; Mohamed, A.S.A. Efficientnet-Lite and Hybrid CNN-KNN Implementation for Facial Expression Recognition on Raspberry Pi. *IEEE Access* **2021**, *9*, 134065–134080. [CrossRef]
47. Deng, J.; Dong, W.; Socher, R.; Li, L.J.; Li, K.; Fei-Fei, L. ImageNet: A large-scale hierarchical image database. In Proceedings of the IEEE Conference on Computer Vision and Pattern Recognition (CVPR), Miami, FL, USA, 20–25 June 2009; pp. 248–255.
48. Sandler, M.; Howard, A.; Zhu, M.; Zhmoginov, A.; Chen, L. MobileNetV2: Inverted Residuals and Linear Bottlenecks. In Proceedings of the 2018 IEEE/CVF Conference on Computer Vision and Pattern Recognition, Salt Lake City, UT, USA, 18–23 June 2018; pp. 4510–4520.
49. Tan, M.; Le, Q.V. EfficientNet: Rethinking Model Scaling for Convolutional Neural Networks. In Proceedings of the 36th International Conference on Machine Learning (ICML), Long Beach, CA, USA, 10–15 June 2019; Volume 97, pp. 6105–6114.

50. Hu, J.; Shen, L.; Sun, G. Squeeze-and-excitation networks. In Proceedings of the IEEE Conference on Computer Vision and Pattern Recognition, Salt Lake City, UT, USA, 18–23 June 2018; pp. 7132–7141.
51. Rajani, N.; McArdle, K.; Dhillon, I.S. Parallel k nearest neighbor graph construction using tree-based data structures. In Proceedings of the 1st High Performance Graph Mining workshop, Sydney, Australia, 10 August 2015; Volume 1, pp. 3–11.
52. Silpa-Anan, C.; Hartley, R. Optimised KD-trees for fast image descriptor matching. In Proceedings of the 2008 IEEE Conference on Computer Vision and Pattern Recognition, Anchorage, AK, USA, 23–28 June 2008; pp. 1–8.
53. Facebook AI Research. Faiss. 2022. Available online: https://github.com/facebookresearch/faiss (accessed on 17 June 2023).
54. Norouzi, M.; Fleet, D.; Salakhutdinov, R. Hamming Distance Metric Learning. In Proceedings of the Advances in Neural Information Processing Systems, Lake Tahoe, NV, USA, 3–8 December 2012; Curran Associates, Inc.: Red Hook, NY, USA, 2012; Volume 25.
55. Pronobis, A.; Caputo, B. COLD: COsy Localization Database. *Int. J. Robot. Res.* **2009**, *28*, 588–594. [CrossRef]
56. Shuvo, M.M.H.; Islam, S.K.; Cheng, J.; Morshed, B.I. Efficient Acceleration of Deep Learning Inference on Resource-Constrained Edge Devices: A Review. *Proc. IEEE* **2023**, *111*, 42–91. [CrossRef]
57. Scaramuzza, D. Omnidirectional Vision: From Calibration to Root Motion Estimation. Ph.D. Thesis, ETH Zurich, Zürich, Switzerland, 2007.
58. Baker, S.; Nayar, S.K. A Theory of Single-Viewpoint Catadioptric Image Formation. *Int. J. Comput. Vis.* **1999**, *35*, 175–196. [CrossRef]
59. Kowa. 4.4-11mm F1.6 LMVZ4411 1/1.8″ Lens. 2023. Available online: https://cmount.com/product/kowa-4-4-11mm-f1-6-lmvz4411-1-1-8-lens-c-mount (accessed on 17 June 2023).
60. Bazin, J.C. Catadioptric Vision for Robotic Applications. Ph.D. Thesis, KAIST, Daejeon, Republic of Korea, 2019.
61. Tan, M.; Le, Q.V. EfficientNet: Rethinking Model Scaling for Convolutional Neural Networks. *arXiv* **2019**, arXiv1905.11946.
62. Tan, M.; Le, Q. EfficientNetV2: Smaller Models and Faster Training. *arXiv* **2021**, arXiv:2104.00298.
63. Nanne. pytorch-NetVlad. 2023. Available online: https://github.com/Nanne/pytorch-NetVlad (accessed on 17 June 2023).
64. Johnson, J.M.; Khoshgoftaar, T.M. Survey on deep learning with class imbalance. *J. Big Data* **2019**, *6*, 27. [CrossRef]

sensors

MDPI

Article

Transparency-Aware Segmentation of Glass Objects to Train RGB-Based Pose Estimators

Maira Weidenbach, Tim Laue and Udo Frese *

Faculty of Mathematics and Computer Science, University of Bremen, 28359 Bremen, Germany; maiweide@uni-bremen.de (M.W.); tlaue@uni-bremen.de (T.L.)
* Correspondence: ufrese@uni-bremen.de

Abstract: Robotic manipulation requires object pose knowledge for the objects of interest. In order to perform typical household chores, a robot needs to be able to estimate 6D poses for objects such as water glasses or salad bowls. This is especially difficult for glass objects, as for these, depth data are mostly disturbed, and in RGB images, occluded objects are still visible. Thus, in this paper, we propose to redefine the ground-truth for training RGB-based pose estimators in two ways: (a) we apply a transparency-aware multisegmentation, in which an image pixel can belong to more than one object, and (b) we use transparency-aware bounding boxes, which always enclose whole objects, even if parts of an object are formally occluded by another object. The latter approach ensures that the size and scale of an object remain more consistent across different images. We train our pose estimator, which was originally designed for opaque objects, with three different ground-truth types on the ClearPose dataset. Just by changing the training data to our transparency-aware segmentation, with no additional glass-specific feature changes in the estimator, the ADD-S AUC value increases by 4.3%. Such a multisegmentation can be created for every dataset that provides a 3D model of the object and its ground-truth pose.

Keywords: neural networks; training data; transparent objects; bounding box; segmentation; pose estimation

Citation: Weidenbach, M.; Laue, T.; Frese, U. Transparency-Aware Segmentation of Glass Objects to Train RGB-Based Pose Estimators. *Sensors* **2024**, *24*, 432. https://doi.org/10.3390/s24020432

Academic Editors: Xin Zhao, Mingzhu Sun and Qili Zhao

Received: 4 December 2023
Revised: 22 December 2023
Accepted: 25 December 2023
Published: 10 January 2024

1. Introduction

Household robotics is a challenging topic as it includes a lot of different tasks in a dynamic environment. For example, robots should be able to set the table, fetch objects, and place them in different places. To be able to manipulate any object, the robot needs to know that the object exists and where it is located in the world. This can be achieved, for example, with cameras perceiving color and depth information of the environment. Many works have been published about 6D pose estimation of opaque objects in different scenarios [1–9]. However, humans tend to use drinking devices, salad bowls, or decanters made out of glass; thus, the robot needs to be able to cope with transparent objects as well as opaque objects. There are two big differences between those objects: depth data are often unreliable, and clutter has a different effect. The problem with the depth data is that normal depth cameras rely on infrared light and laser range finders use high-frequency light. This may work fine with opaque objects where the light can be reflected but not really with transparent objects, as light just passes through and is either lost or reflected by objects or walls somewhere behind the actual object. This can distort the depth data quite a lot. There are different common approaches to tackle this problem: trying to reconstruct the depth data, as performed for example in [10–12], simply trying to solve the 6D pose estimation with stereo RGB images [13] or by just using single RGB images [14,15]. The latter is what our work focuses on.

Another difference between opaque and transparent objects is that clutter might not have the same visual effect. If an opaque object stands in front of another object, it covers the other object to a certain degree, and the object behind is only partly visible. With transparent objects, however, this is not always the case. A big glass object in front of another usually only covers small parts on its edges where the light reflects differently. But because its main part is just transparent, it is possible to see the object behind it quite well. Figure 1 highlights that the big salad bowl on the table (Figure 1a) stands in front of four other objects, but those are still visible, as a closer look on the wine glass in Figure 1c shows. Nevertheless, the ground-truth segmentation and the resulting bounding box Figure 1b are, by definition, only on the directly visible part of the object. This adds another difficulty for two-stage approaches. Usually, they first detect the object within a bounding box and scale the bounding box to a fixed size to then run a second network on that image. Now, if an object is largely occluded, the remaining part is scaled (Figure 1b), and consequently, the object has a much different scale than if not occluded. It might be quite helpful to prevent the wrong scaling and extend the bounding box to the whole object (Figure 1c) and teach the network that the hidden but visible part also belongs to the object. Thus, we propose to rethink the bounding box and segmentation for transparent clutter. We call this transparency-aware segmentation (TAS) and transparency-aware bounding box (TAB). The contributions of this paper are to

- modify ground-truth segmentation and bounding boxes to cover whole objects in cluttered transparent scenes,
- compare training with transparency-aware and original segmentation and its influence on the pose,
- apply our approach to a cluttered challenging glass scene using only RGB data without modifications specific to the glass dataset.

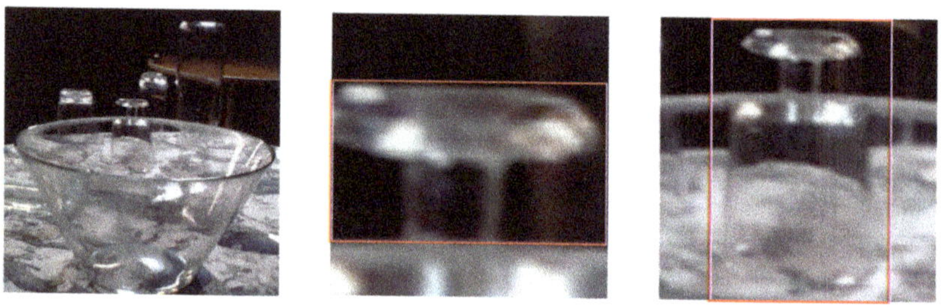

(**a**) (**b**) (**c**)

Figure 1. Example crops from a scene in the ClearPose [16] dataset with typical glass clutter, a wide overview (**a**) of the scene. The original bounding box (**b**) and an object transparency-aware bounding box (**c**), as well as their resulting regions of interest for the second stage.

2. Related Work

As knowing the 6D pose of an object is probably essential for any robotic manipulation, there exist various methods to estimate it. They can be differentiated between direct methods that directly output a 6D pose such as PoseNet [17] and GDR-Net (geometry-guided direct regression) [18], or indirect methods that produce intermediate outputs and use RANSAC or PnP solvers to estimate the pose [1,15]. Another conceptual difference is the number of input images. Zebrapose [1] is a single-image approach that involves splitting the object into multiple regions with several sub-regions, interpreting every region as a certain class and as a multi-class classification problem before solving the resulting correspondences via a PnP solver. A multiple-view approach is, e.g., CosyPose [6], which uses a single image for initial object candidates, matches those candidates across other views, and globally refines object and camera poses. Another approach improving most results is to include depth data. FFB6D (full-flow bidirectional fusion network for 6D pose estimation) continuously fuses information from RGB and depth data within their encoder and decoder structured networks, followed by a 3D key-point detection and instance segmentation and least-square fitting [3].

All those methods, as well as most other work in this area, focus on texture-less, cluttered, and opaque objects; however, only a few works have been published specifically on transparent objects. As pointed out before, transparent objects face some further difficulties. There are some approaches using depth data, with some work on depth completion, as in [10], which uses affordance maps to reconstruct the depth data. Another approach is made by Xu et al. [11]. They try to reconstruct the depth image with by surface normal recovery from the RGB image and use this to calculate an extended point cloud. Using the RGB image and the constructed point cloud, another network estimates the final pose. Because of this depth problem there are also direct methods to use only RGB images like [14]. They adapt the idea of GDR-Net to use geometric features of objects and fuse those with edge features and surface fragments specified for transparent objects. From these fused features, a dense coordinate and confidence map is obtained and processed by GDR's patch-PnP. An indirect approach is [15] by Byambaa et al. They also take only a single RGB image, but apply a two-stage approach where a neural network extracts 2D key points, followed by a PnP algorithm afterwards. These works have in common that they adapt their networks somehow to the specific characteristics of transparent materials, making them more specific and less suitable for general-purpose object pose estimation. We want to keep our network and training structure identical to previous work and rather change the ground-truth definition of the training data itself in a glass-dependent way. The idea is to enhance pose estimation on transparent objects without reducing the generality of the estimator.

In previous work, our group has proposed an indirect general approach for RGB and RGB-D images to estimate 6D poses [4] with a specific focus on symmetrical objects [19]. It splits the process into two stages, based on the idea that neural networks have been proven to predict well which object is where in the image and that mathematical algorithms can solve 2D-3D correspondence problems. In the first stage, a neural network is trained to predict a segmentation, an object image, and an uncertainty map of the predicted object image. This object image represents in each pixel the corresponding 3D point of the object in object coordinates in a second stage. These learned correspondences of the 2D pixels with the 3D object points within the object image are then solved by a generalized PnP algorithm, taking the uncertainty into account. This approach works well on symmetrical and opaque objects. In contrast to Xu et al., we do not change the structural training network and extend it with specific features for glass but rather reuse our general approach and simply adjust the training data to achieve better results on transparent objects.

Definitions for bounding boxes and segmentation of objects, which are necessary for most training processes in the area of 6D pose estimation, are hardly mentioned and discussed. Sometimes, it is implicitly displayed in figures [14] or encoded in the choice of the object detector [1] but rarely explicitly declared or discussed. In the case of bounding boxes, most works estimate axis-aligned bounding boxes, as in [20–22], and only few estimate object-oriented bounding boxes [23,24], but to our knowledge, there is no discussion about segmentation and the fact that one pixel can belong to more than one object, e.g., in the case of transparent objects or scenes seen through a window.

3. Approach

We want to investigate whether changing the training data considering the specialities of transparent objects enhances the pose estimation without having to adapt the network with special glass features. To achieve this, we use a multisegmentation. Conceptually speaking, we extend the normal segmentation mask, where each pixel is assigned to exactly one object or background, by assigning multiple objects to pixels in cases where clutter normally hides object pixels. The idea is visualized in Figure 2b. The implementation is performed as one segmentation image for each object individually within the image. Modifying and testing different training data with our indirectly working pose estimator requires different processes for training and prediction.

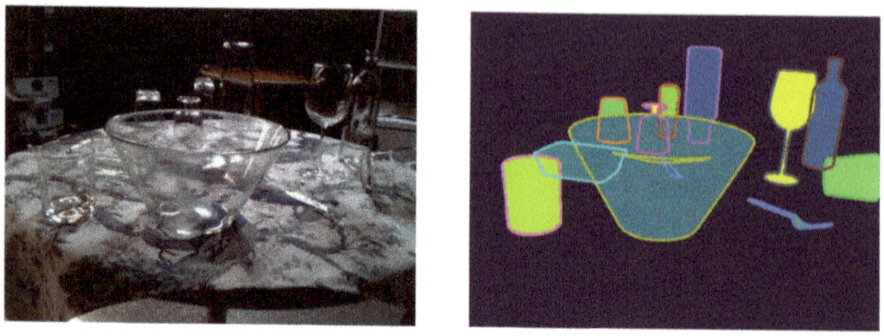

(**a**) (**b**)

Figure 2. Example RGB image of the dataset (**a**) and the corresponding multisegmentation mask (**b**), where the fully colored pixels represent the original ground-truth segmentation, and the object border indicates that the objects continue. The pixels within the borders belong to all overlapping objects, not just to the object in front.

3.1. Training

In the training processes, sketched in Figure 3, we consider three different variants in the segmentation and bounding box process. We want to investigate and assess the effect of a consistent size and scale of the object (transparency-aware bounding box), as well as of the transparency-aware segmentation definition. Figure 4 visualizes the three differences. In the first variant (Figure 4a), the original ground-truth mask and the resulting bounding box are used. The second variant (Figure 4b) also uses the ground-truth mask, but the transparency-aware bounding box (TAB) encloses the whole object plus an additional five pixels to every side. This ensures a consistent size and scale of the object and adds some more context around the object. The third training set is built by using the transparency-aware segmentation (TAS) mask and bounding box (Figure 4c), including even the formally occluded parts of the object. As glass is transparent, i.e., even though there might be some glass in front of the object of interest, the object of interest is still partly visible. Only the edges of the occluding objects really disturb the image. Having a fixed quadratic input size of the networks requires using the bounding boxes and calculating a region of interest that fits the networks' needs. As in our previous work [4], we use the longest side of the bounding box to fit a square around the object (uniform scaling). The resulting ROI is used to crop the segmentation, RGB image, and our ground-truth object image. The images are either upscaled or downscaled to meet the second stage input size. Then, the two networks are trained independently of each other.

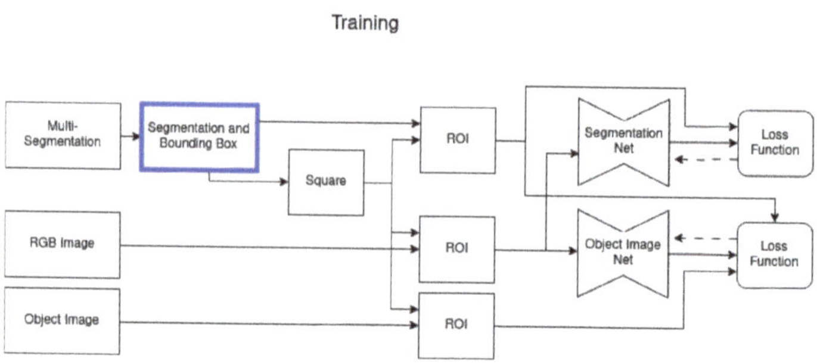

Figure 3. In the training phase, a multisegmentation is used to create our novel segmentation masks and bounding boxes, which is our main contribution (blue box). The dimensions of the bounding box are used to derive a squared region of interest. This region is used to crop the RGB input image as well as the ground-truth object image and segmentation. The two networks are trained independently, the segmentation network is trained to predict a segmentation map based on the defined ground truth, and the object image network is trained on the ground-truth object image. As its loss is on a pixel level, the ground-truth segmentation mask is used to only account for the loss for the pixels that are actually part of the object.

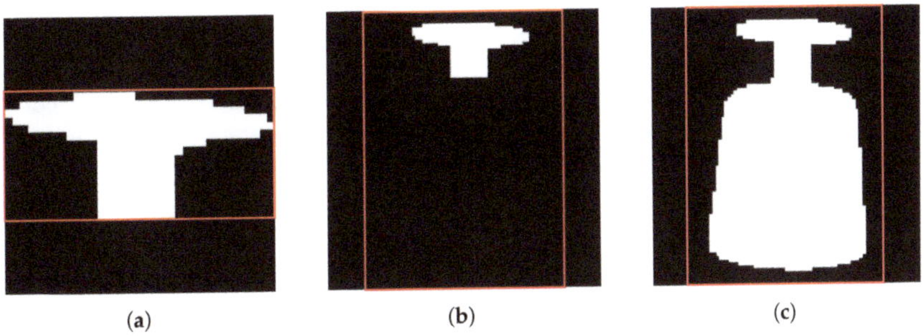

Figure 4. Different mask and bounding box sizes for the RGB image crops of Figure 1. (**a**) Displays the original mask and resulting bounding box, and (**b**) also uses the original mask but the transparency-aware bounding box. The right image (**c**) displays the transparency-aware segmentation and its resulting bounding box.

3.2. Prediction

At prediction time, an external set of bounding boxes of specific objects of the complete image is needed. To obtain these bounding boxes, any object detection algorithm can be used. This bounding box detection is a slightly different problem and out of the scope of this paper. Many methods were developed to find and classify specific objects within an image and produce bounding boxes [20–22,25,26]. These boxes can then be used to find more attributes like the color, materials, a text description of the object, and the 6D pose of the object. We focus on 6D pose estimation. Thus, for the prediction phase in our approach, we use the ground-truth boxes and the transparency-aware boxes, respectively, for the network training. For a real-life application, any object detector could be connected in front of our workflow, which is visualized in Figure 5.

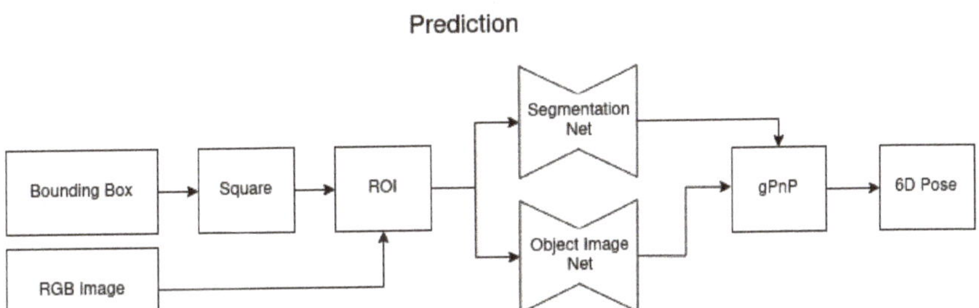

Figure 5. In the prediction phase, a bounding box for the object of interest is needed. It can come from any object detector. The dimensions of the bounding box are used to create a square around the object and use this as a region of interest for further processing. The incoming RGB image is cropped and scaled by this ROI and fed into a pre-trained segmentation and object image network. The object image network estimates the 3D points in object coordinates and an uncertainty map in pixel space. These two are used pixel-wise to solve the 2D-3D correspondence with the gPnP, while only the pixels are taken into account that belong to the object estimated by the segmentation network. This resolves into a 6D pose for this object.

4. Dataset

We chose the ClearPose dataset [16] to test our approach because it contains several real image scenes, in total over 350,000 annotated images, with many transparent, symmetric objects within cluttered scenes and annotated ground-truth segmentation maps and ground-truth poses. Other known available real-world datasets are ClearGrasp [27] and TransCG [28], which use plastic glasses instead of realistic real glass and include almost no glass clutter, making them uninteresting for our comparison. The complete ClearPose dataset includes several sets of scenes of different types, for example, normal objects placed jumbled on a table but also objects with filled liquors or with an additional translucent cover. We focus on set number 4. Set 4 includes about 43,000 RGB-D images with segmentation masks and poses. It consists of 6 scenes, all of them including the same 12 different objects: drinking glasses, bottles, bowls, and a knife and a fork, either made of glass or plastic, placed on a table. The difference between the scenes is their colorful tablecloth and the arrangement of the objects. This arrangement is random but physically correct, i.e., glasses are standing upright or upside down or lying flat on the table but do not balance on edges or lean on each other. The lighting seems to be realistic indoor room light, produced by fluorescent tubes and spotlights. To obtain the different images, a camera was moved around the table at different heights. This results in different light reflections and blurriness in the images. Sometimes, the unstructured environment of the lab where the images were taken is visible, depending on the camera position.

The ground-truth data were generated by a digital twin scene of the real scene and an estimated camera movement using ORB-SLAM3. With this information, the ground-truth segmentation and the true-depth image were rendered. This makes total sense for opaque objects but not so much for transparent objects. As visible in Figure 1, some pixels show different objects at the same time, where the bowl and the wine glass are behind it. The ground-truth segmentation is a gray image, with only one channel, encoding for each pixel if it is background or belongs to a specific object, but it does not represent multiple objects at the same pixel. Thus, for our experiment, we extended the ground-truth segmentation mask by using the provided 6D pose of the object and the provided 3D model and rendered a transparency-aware segmentation for each object in each image. For training, we used the included ground-truth and our own multisegmentation. Figure 2 shows one example image of the test set and our rendered multisegmentation.

5. Experiments

To evaluate our idea, we built expert networks for each object in Set 4, meaning we have one network for each object. In theory, this enables us to be unlimited in the number of objects we can process. However, in practice, this is limited by the memory needed. As we use these networks to evaluate Set 4 of the ClearPose dataset, we have 12 networks, one for each object, which we evaluate sequentially.

5.1. Network Architecture

Our network is not new, but for the sake of completeness, we would like to present the used network architecture. In contrast to our approach in [19], we separate the original network into two networks for the segmentation and for the object image and train both of them independently. Although both have the same structural encoder (Table 1) with three layers of a pre-trained MobileNet and additional convolution layers, the decoder (Table 2) is different. This is schematically shown in Figure 6. The segmentation network predicts the segmentation, and the object image network predicts the object image, as well as an uncertainty map representing the uncertainty of each pixel in image space. The object image network from Figure 3 is in detail divided into one encoder and three heads as sketched in Figure 6. Table 2 shows the used structure of layers for each head, where they only differ in their output layer.

5.2. Training

We trained our networks on three NVIDIA A40 and two NVIDIA TITAN V using TensorFlow 2.11 and Python 3.8. To train the segmentation networks, we used a batch size of 64 and Adam as the optimizer with a fixed learning rate of 0.0001 and trained for 300 epochs. The object image network was trained with a batch size of 128 and used Adam as the optimizer with a fixed learning rate of 0.001 for 100 epochs on the object image and for another 200 epochs on the uncertainty map. The differences between the learning parameters resulted from the different computation powers. For each training set, we trained both networks for all objects, keeping all structural parameters the same, with no extra modification for any object or the training set. The networks have an input size of 224 × 224 and produce a segmentation mask, an object image, and an uncertainty map, each of size 112 × 112. Training and evaluation are performed on these three different training sets for each individual object to investigate the differences in 6D pose estimation. For a real-life application, an additional bounding box detector would be necessary.

As a metric, we use a symmetry-aware average point distance error (ADD-S) because almost all objects are somehow symmetrical. Only the knife and fork do not have a rotation symmetry axis. This is based on the pairwise average distance (ADD) [29] but includes the fact that points are ambiguous. Pure ADD would require the network to learn to guess the annotated rotation, which is arbitrarily set by the annotation. ADD-S is suitable to handle symmetrical and asymmetrical objects. We calculate the area under the curve (AUC) for thresholds from 0.0 m to 0.1 m, as in [16].

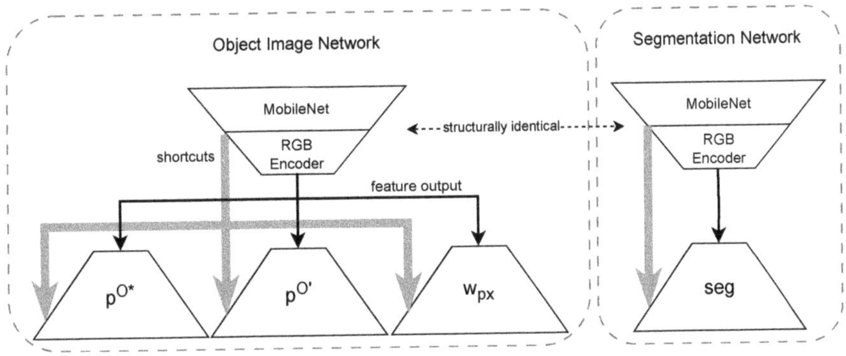

Figure 6. A more detailed sketch of our used network structure. The object image network has three heads. p^{O*} and $p^{O'}$ represent two different aspects of the object image points p^O, which is described in more detail in [19]. As this is just a technical detail, we will refer to p^{O*} and $p^{O'}$ together as an object image in the following. The pixel weight w_{px} estimates the uncertainty in pixel space. In the first training stage, only the object images (p^{O*} and $p^{O'}$) are trained. In the second stage, the weights of the encoder are set and only the pixel weight head (w_{px}) is trained. The segmentation network has a simple common encoder–decoder with a shortcut structure. The layers of the encoder and decoder are listed in Tables 1 and 2.

Table 1. RGB encoder structure.

Layer Name	Nr. of Output Channels	Filter Size	Stride	Output Size	Nr. of Layers
MN block_3_project_BN	32	-	-	28 × 28	-
Conv1	100	5 × 5	2	14 × 14	1
Dense1	64	-	1	14 × 14	12 × 3
Conv2	128	1 × 1	1	14 × 14	1
Dense2	64	-	1	14 × 14	12 × 3
Conv3	128	1 × 1	1	14 × 14	1
Conv4	70	5 × 5	2	7 × 7	1
Dense3	70	-	1	7 × 7	6 × 3
Conv5	140	1 × 1	1	7 × 7	1

MN block_3_project_BN—the used output layer of the MobileNet v2 version of keras, ConvX—a 2D convolution operation, DenseX—a custom dense block as described in Table 3.

Table 2. Decoder structure.

Layer Name	Nr. of Output Channels	Filter Size	Stride	Output Size	Nr. of Layers
Up1	140	-	-	14 × 14	1
Concat2	268	Conv3 (Enc.) & Conv5 (Enc.)		14 × 14	1
Conv6	50	3 × 3	1	14 × 14	1
Dense4	25	-	1	14 × 14	12 × 3
Conv7	50	1 × 1	1	14 × 14	1
Up2	50	-	-	28 × 28	1
Concat3	82	Dense4 & MN block_3_project_BN		28 × 28	1
Conv8	32	3 × 3	1	28 × 28	1
Dense5	16	-	1	28 × 28	6 × 3
Conv9	32	1 × 1	1	28 × 28	1
Up3	32	-	-	56 × 56	1
Concat4	56	Dense5 & MN block_2_project_BN		56 × 56	1
Conv10	24	3 × 3	1	56 × 56	1
Dense6	12	-	1	56 × 56	4 × 3
Conv11	24	1 × 1	1	56 × 56	1
Up4	24	-	-	112 × 112	1
Concat5	56	Dense6 & MN bn_Conv1		112 × 112	1
Conv12	16	3 × 3	1	112 × 112	1
Dense7	8	-	1	112 × 112	4 × 3
Conv13	16	1 × 1	1	112 × 112	1
Conv ($p^{O*}/p^{O'}/w_{px}$/seg)	3/3/4/1	1 × 1	1	112 × 112	1

Enc.—Layer from the Encoder, ConvX—a 2D convolution operation, ConcatX—a concatenation layer, UpX a 2D upscaling layer, DenseX—a custom dense block as described in Table 3, MN X—the used layer of the MobileNet v2 version of keras.

Table 3. Dense block structure.

Layer Name	Nr. of Output Channels	Filter Size	Stride	Output Size	Nr. of Layers
ConvI	input channels	1 × 1	1	input size	1
ConvII	input channels	3 × 3	1	input size	1
Concat		Input Layer & ConvII		input size	1

ConvX—a 2D convolution operation, ConcatX—a concatenation layer. The number of channels and the output size depends on the layer before, where the input channels and input size is determined.

6. Results

We trained both networks separately. The segmentation network can not only give insights on their individual performance but also whether the network structure is suitable to identify the object at all. As stated before, the encoder structure of both networks is identical, and only the decoder differs.

6.1. Segmentation

The segmentation networks were trained for each object individually, each on the three different training sets: original, transparency-aware bounding box (TAB), and transparency-aware segmentation including the bounding box (TAS). Table 4 displays the recall and precision values on a pixel level averaged over all objects. The highest values were achieved by our transparency-aware segmentation. However, it is interesting that the transparency-aware bounding boxes have the lowest values. This might be because, for the additional context, without TAS, a lot more background pixels are added, producing a slight bias to background pixels as the precision is similar, but the recall values are worse. Overall, this also means that our network structure is, in general, suitable to detect object pixels within the image. Otherwise, the segmentation would not be as good.

Table 4. Results of the test set segmentation on a pixel level. Best results in bold.

	Original	**TAB**	**TAS**
recall	83.64%	79.13%	**89.53%**
precision	93.10%	93.82%	**97.22%**

6.2. 6D Pose Estimation

For the pose estimation, we use the prediction stage as displayed in Figure 5. The bounding boxes are either based on the original segmentation or the transparency-aware segmentation, depending on what data the respective network was trained on. The overall results on Set 4 are displayed in Table 5. Changing the ground-truth definition from original to TAS increases the AUC by 4.33% over all objects.

In order to compare our results with a reference value on this glass data set and to put them into a context, we compared them to Xu et al. [11] and FFB6D [3]. However, we do not claim to compete against their full approaches, as we use ground-truth bounding boxes without using any object detector. Even though [14,15] are methodologically more similar to our approach, we cannot compare ourselves to them as they use different data sets with little to no occlusion. Xu et al. developed a network with specific glass features, and FFB6D was a state-of-the-art network for pose estimation at the time when the dataset was published. Both have better results; however, both use RGB-D data, whereas our network only uses RGB data. In previous work, including depth data achieved an increase of up to 24% percent [4]. When FFB6D was trained and tested with ground-truth depth data, it achieved an improvement of almost 15%.

To highlight the difficulty of the dataset, we show two example images and the resulting pose in Figure 7. The used networks were trained with TAS.

We evaluated our networks sequentially and a 6D estimation prediction step took 0.175 s on average. This includes the segmentation and object image network as well as the complete gPnP stage for all objects in the image. However, this excludes the time needed to load the networks, which would be necessary if the RAM is not big enough to keep all networks needed in memory. In our experiments, we needed around 42GB RAM to have all 12 networks running at the same time. It is worth noting that an additional object detector would also need some resources in a real-world application.

To distinguish and have a more fine-grained result, in Figure 8 we calculated the ADD-S accuracy-threshold curve for all objects up to a threshold of 0.1 m, which is the same as the threshold for the ADD-S AUC. Interestingly, the results are quite different for different objects. It seems to work best on tall objects with a continuous symmetry axis. Not surprising is the fact that the knife and fork do not perform well. The special properties of these two objects are discussed in detail in Section 6.4. As our network could not deal with these two objects, we omit them in the next evaluation step.

Table 5. Results of the test set. The results from Xu et al. and FFB6D are taken from [16].

	Xu et al. [11]	**FFB6D** [3]	**Original**	**TAB**	**TAS**
ADD-S AUC	45.23%	43.45%	36.27%	38.27%	40.60%
without knife+fork	-	-	43.26%	45.85%	48.36%

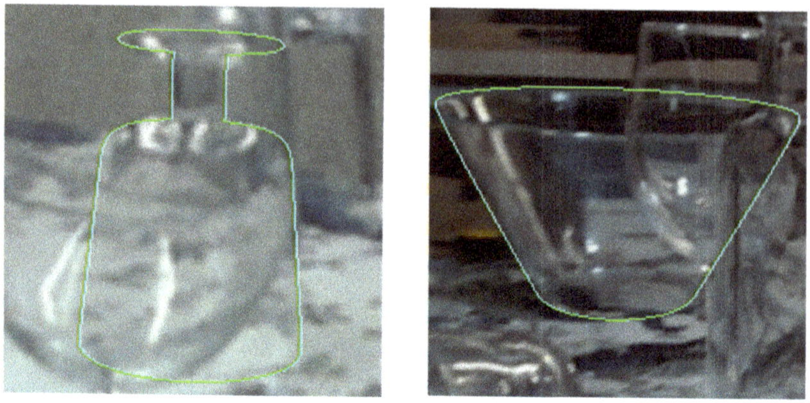

(**a**) (**b**)

Figure 7. Two examples of difficult images where the estimated pose is rendered as a green contour on top of the RGB image. In (**a**), the wine glass is almost completely covered by another wine glass with reflections from the fluorescent tubes and has another glass in the back which also causes some edge distractions. The salad bowl (**b**) has multiple objects in the front and much background clutter.

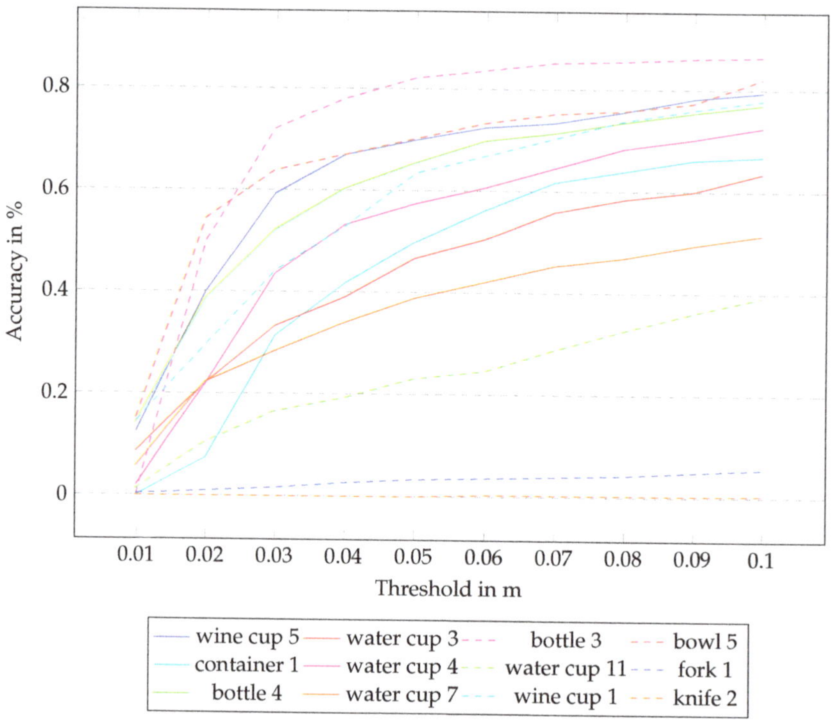

Figure 8. ADD-S accuracy-threshold curve of all 12 objects in Set 4 trained by using TAS.

6.3. *Dependence on Level of Occlusion*

To investigate the effect of the different training data, we performed an ablation study comparing the original bounding box and segmentation approach to our TAB and TAS. To obtain a more fine-grained result in relation to covered glass, we split the test data by their visibility using the percentage of the number of originally assigned ground-truth pixels to the number of TAS-rendered segmentation pixels (Figure 9). It is noticeable that our transparency-aware segmentation increases the AUC by 20% in the range of visibility up to 40% compared to the original. The strong effect leverages out as more of the object becomes visible, which makes sense because the training data is becoming similar. It is also noteworthy that the TAB also increases the AUC up to 14% in the range of 20–40% visibility and also performs better on all other images compared to the original. It seems like having the right object size and scale helps obtain a good object image, which encodes the geometric form, because even though the segmentation is worse than in the original approach, the 6D pose is significantly better, up to a visibility of 80%. This effect vanishes in the last scenario with 80–100% visibility. The difference in the aspect ratios of the two versions is not that big anymore as the original also encloses the whole object. The only real difference is the additional context. This was created by adding five pixels in each direction. The results indicate that this has only little influence as original and TAB are almost equivalent. An interesting effect occurs in the area 40–80%. The TAB achieves higher values than our TAS training data. Manual insights into test data in this range (see four examples in Figure 10) indicate that in these images, there is a lot of overlapping, disturbing the view of the object behind, as the edges reflect the light differently. This can result in actual hiding as in Figure 10a, where the right part is just not visible anymore, or in deforming the object as in Figure 10b, where the bottom of the bowl visually bends as seen through the glass in front of it. In these cases, it seems logical that the TAS might be too big, covering parts that do not belong to the object visually. It still achieves better results than the original training data.

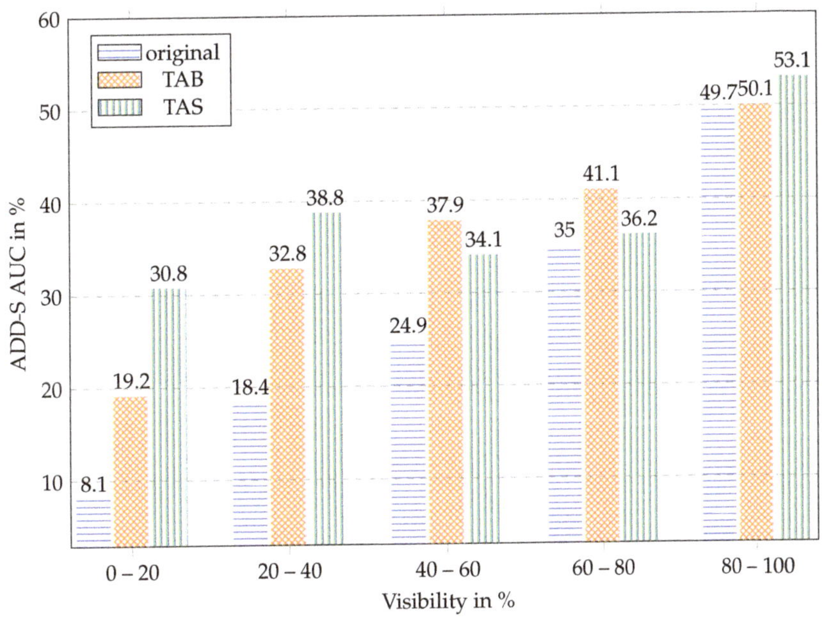

Figure 9. Results of the different training data distributed over the visibility of the objects, e.g., the bar of 0–20% visibility includes all images where the object is, based on the ground truth, only up to 20% visible.

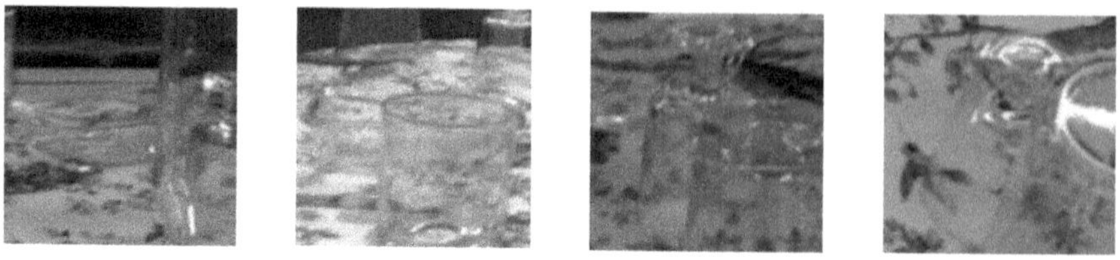

(a) (b) (c) (d)

Figure 10. Example images of one bowl and wine glass with different visibility values around 50–70%: (**a**) bowl, 71% visibility; (**b**) bowl, 54% visibility; (**c**) wine glass, 55% visibility; (**d**) wine glass, 66% visibility.

6.4. Knife and Fork

There are two very challenging objects within the dataset: a transparent knife and fork (see examples in Figure 11). Both objects were hardly recognized by our network and did not profit from the different training settings. These two objects probably profit the most from the depth data, as they always lay flat on the table. Even if the depth camera does not recognize the object, it will detect the table underneath it. The fork has a maximum height of 15 mm and a knife of about 2 mm, meaning that the table depth is still a good approximation of the depth data for these objects. But without depth data, in RGB-only networks like ours, these objects could be anywhere in the room, making it really hard to predict. If we exclude these two objects, we obtain an AUC value of 43.26% with the original input and 48.36% for our transparency-aware approach.

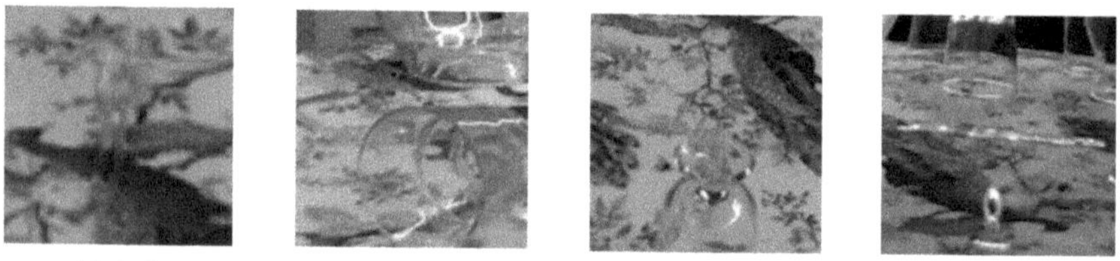

(**a**) Fork (**b**) Fork (**c**) Knife (**d**) Knife

Figure 11. Example images of the fork (**a**,**b**) and the knife (**c**,**d**). Even if only the object is present, no distractions, (**a**), it can be really hard to see the fork (it is centered and points upwards). In (**b**,**c**) other glasses are within the bounding box and it is almost impossible to see the objects of interest. The right image highlights that even if the knife is visible, it has a completely different appearance because of reflections along the whole object.

7. Conclusions

We introduced a new way of defining segmentation masks and bounding boxes for transparent objects, being aware of the characteristics of glass and other transparent materials, to improve RGB-based pose estimation without having to adapt the network itself. With only a little effort, we render the multisegmentation mask, which can be used to create transparency-aware segmentation and bounding boxes. This can be achieved independently of the dataset if 3D object models and ground-truth poses are available. Using these instead of the original training data in the training process benefits the 6D pose estimation of a given RGB estimator if the objects can be recognized by the network at all. This is particularly true when there is strong occlusion and objects are only partially visible according to the original definition.

One downside of our approach is that a 3D object model and a ground-truth pose have to exist to generate the transparency-aware segmentation mask. This is not as easily applicable if the objects of interest are parameterizable or articulated without a fixed model. However, most work is still required on fixed and stable objects; thus, it seems like a good and easy option to increase the accuracy of 6D pose estimation for transparent objects just by adapting the training set. The success of TAB suggests the hypothesis that presenting the objects in the right scale, even if they are occluded in the training data, as it is with the original segmentation, might also help 6D pose estimation on opaque objects.

Author Contributions: Conceptualization, M.W., T.L. and U.F.; methodology, M.W., T.L. and U.F.; software, M.W.; validation, M.W.; formal analysis, M.W., T.L. and U.F.; investigation, M.W.; resources, M.W.; data curation, M.W.; writing—original draft preparation, M.W.; writing—review and editing, T.L. and U.F.; visualization, M.W.; supervision, T.L. and U.F.; project administration, U.F.; funding acquisition, U.F. All authors have read and agreed to the published version of the manuscript.

Funding: The research reported in this paper has been supported by the German Research Foundation DFG, as part of Collaborative Research Center 1320 EASE—Everyday Activity Science and Engineering. https://ease-crc.org/ (accessed on 24 December 2023).

Institutional Review Board Statement: Not applicable.

Informed Consent Statement: Not applicable.

Data Availability Statement: Publicly available datasets were analyzed in this study. This data can be found here: https://github.com/opipari/ClearPose accessed on: 24 December 2023.

Conflicts of Interest: The authors declare no conflicts of interest. The funders had no role in the design of the study; in the collection, analyses, or interpretation of data; in the writing of the manuscript; or in the decision to publish the results.

Abbreviations

The following abbreviations are used in this manuscript:

ADD-S	Symmetry-aware average point distance
AUC	Area under the curve
TAB	Transparency-aware bounding box
TAS	Transparency-aware segmentation
ROI	Region of interest

References

1. Su, Y.; Saleh, M.; Fetzer, T.; Rambach, J.; Navab, N.; Busam, B.; Stricker, D.; Tombari, F. ZebraPose: Coarse to Fine Surface Encoding for 6DoF Object Pose Estimation. *arXiv* **2022**, arXiv:2203.09418.
2. Park, K.; Patten, T.; Vincze, M. Pix2Pose: Pixel-Wise Coordinate Regression of Objects for 6D Pose Estimation. In Proceedings of the IEEE/CVF International Conference on Computer Vision (ICCV), Seoul, Republic of Korea, 27 October–2 November 2019.
3. He, Y.; Huang, H.; Fan, H.; Chen, Q.; Sun, J. FFB6D: A Full Flow Bidirectional Fusion Network for 6D Pose Estimation. *arXiv* **2021**, arXiv:2103.02242.
4. Richter-Klug, J.; Frese, U. Towards Meaningful Uncertainty Information for CNN Based 6D Pose Estimates. In Proceedings of the Computer Vision Systems; Tzovaras, D., Giakoumis, D., Vincze, M., Argyros, A., Eds.; Springer: Cham, Switzerland, 2019; pp. 408–422.
5. Wang, C.; Xu, D.; Zhu, Y.; Martin-Martin, R.; Lu, C.; Fei-Fei, L.; Savarese, S. DenseFusion: 6D Object Pose Estimation by Iterative Dense Fusion. In Proceedings of the IEEE/CVF Conference on Computer Vision and Pattern Recognition (CVPR), Long Beach, CA, USA, 15–20 June 2019.
6. Labbé, Y.; Carpentier, J.; Aubry, M.; Sivic, J. CosyPose: Consistent multi-view multi-object 6D pose estimation. *arXiv* **2020**, arXiv:2008.08465.
7. Li, Y.; Wang, G.; Ji, X.; Xiang, Y.; Fox, D. DeepIM: Deep Iterative Matching for 6D Pose Estimation. *Int. J. Comput. Vis.* **2019**, *128*, 657–678. [CrossRef]
8. Krull, A.; Brachmann, E.; Nowozin, S.; Michel, F.; Shotton, J.; Rother, C. PoseAgent: Budget-Constrained 6D Object Pose Estimation via Reinforcement Learning. In Proceedings of the IEEE Conference on Computer Vision and Pattern Recognition (CVPR), Honolulu, HI, USA, 21–26 July 2017.
9. Rad, M.; Lepetit, V. BB8: A Scalable, Accurate, Robust to Partial Occlusion Method for Predicting the 3D Poses of Challenging Objects Without Using Depth. In Proceedings of the IEEE International Conference on Computer Vision (ICCV), Venice, Italy, 22–29 October 2017.
10. Jiang, J.; Cao, G.; Do, T.T.; Luo, S. A4T: Hierarchical affordance detection for transparent objects depth reconstruction and manipulation. *IEEE Robot. Autom. Lett.* **2022**, *7*, 9826–9833. [CrossRef]

11. Xu, C.; Chen, J.; Yao, M.; Zhou, J.; Zhang, L.; Liu, Y. 6DoF Pose Estimation of Transparent Object from a Single RGB-D Image. *Sensors* **2020**, *20*, 6790. [CrossRef] [PubMed]
12. Tang, Y.; Chen, J.; Yang, Z.; Lin, Z.; Li, Q.; Liu, W. DepthGrasp: Depth Completion of Transparent Objects Using Self-Attentive Adversarial Network with Spectral Residual for Grasping. In Proceedings of the 2021 IEEE/RSJ International Conference on Intelligent Robots and Systems (IROS), Prague, Czech Republic, 27 September–1 October 2021; pp. 5710–5716. [CrossRef]
13. Liu, X.; Jonschkowski, R.; Angelova, A.; Konolige, K. Keypose: Multi-view 3d labeling and keypoint estimation for transparent objects. In Proceedings of the IEEE/CVF Conference on Computer Vision and Pattern Recognition, Seattle, WA, USA, 14–19 June 2020; pp. 11602–11610.
14. Yu, H.; Li, S.; Liu, H.; Xia, C.; Ding, W.; Liang, B. TGF-Net: Sim2Real Transparent Object 6D Pose Estimation Based on Geometric Fusion. *IEEE Robot. Autom. Lett.* **2023**, *8*, 3868–3875. [CrossRef]
15. Byambaa, M.; Koutaki, G.; Choimaa, L. 6D Pose Estimation of Transparent Object From Single RGB Image for Robotic Manipulation. *IEEE Access* **2022**, *10*, 114897–114906. [CrossRef]
16. Chen, X.; Zhang, H.; Yu, Z.; Opipari, A.; Jenkins, O.C. ClearPose: Large-scale Transparent Object Dataset and Benchmark. *arXiv* **2022**, arXiv:2203.03890.
17. Kendall, A.; Grimes, M.; Cipolla, R. PoseNet: A Convolutional Network for Real-Time 6-DOF Camera Relocalization. *arXiv* **2016**, arXiv:1505.07427.
18. Wang, G.; Manhardt, F.; Tombari, F.; Ji, X. GDR-Net: Geometry-Guided Direct Regression Network for Monocular 6D Object Pose Estimation. *arXiv* **2021**, arXiv:2102.12145.
19. Richter-Klug, J.; Frese, U. Handling Object Symmetries in CNN-based Pose Estimation. In Proceedings of the 2021 IEEE International Conference on Robotics and Automation (ICRA), Xi'an, China, 30 May–5 June 2021; pp. 13850–13856. [CrossRef]
20. Redmon, J.; Divvala, S.; Girshick, R.; Farhadi, A. You Only Look Once: Unified, Real-Time Object Detection. In Proceedings of the 2016 IEEE Conference on Computer Vision and Pattern Recognition (CVPR), Las Vegas, NV, USA, 27–30 June 2016; pp. 779–788. [CrossRef]
21. Ren, S.; He, K.; Girshick, R.; Sun, J. Faster R-CNN: Towards Real-Time Object Detection with Region Proposal Networks. *arXiv* **2016**, arXiv:1506.01497.
22. Tan, M.; Pang, R.; Le, Q.V. EfficientDet: Scalable and Efficient Object Detection. In Proceedings of the IEEE/CVF Conference on Computer Vision and Pattern Recognition (CVPR), Seattle, WA, USA, 14–19 June 2020.
23. Zand, M.; Etemad, A.; Greenspan, M. Oriented Bounding Boxes for Small and Freely Rotated Objects. *IEEE Trans. Geosci. Remote Sens.* **2022**, *60*, 1–15. [CrossRef]
24. Yang, X.; Yang, J.; Yan, J.; Zhang, Y.; Zhang, T.; Guo, Z.; Xian, S.; Fu, K. SCRDet: Towards More Robust Detection for Small, Cluttered and Rotated Objects. *arXiv* **2019**, arXiv:1811.07126.
25. Duan, K.; Bai, S.; Xie, L.; Qi, H.; Huang, Q.; Tian, Q. CenterNet: Keypoint Triplets for Object Detection. In Proceedings of the IEEE/CVF International Conference on Computer Vision (ICCV), Seoul, Republic of Korea, 27 October–2 November 2019.
26. Yoo, D.; Park, S.; Lee, J.Y.; Paek, A.S.; Kweon, I.S. AttentionNet: Aggregating Weak Directions for Accurate Object Detection. In Proceedings of the 2015 IEEE International Conference on Computer Vision (ICCV), Santiago, Chile, 7–13 December 2015; pp. 2659–2667. [CrossRef]
27. Sajjan, S.S.; Moore, M.; Pan, M.; Nagaraja, G.; Lee, J.; Zeng, A.; Song, S. ClearGrasp: 3D Shape Estimation of Transparent Objects for Manipulation. *arXiv* **2019**, arXiv:1910.02550.
28. Fang, H.; Fang, H.S.; Xu, S.; Lu, C. TransCG: A Large-Scale Real-World Dataset for Transparent Object Depth Completion and a Grasping Baseline. *IEEE Robot. Autom. Lett.* **2022**, *7*, 7383–7390. [CrossRef]
29. Xiang, Y.; Schmidt, T.; Narayanan, V.; Fox, D. PoseCNN: A Convolutional Neural Network for 6D Object Pose Estimation in Cluttered Scenes. *arXiv* **2018**, arXiv:1711.00199.

MDPI AG
Grosspeteranlage 5
4052 Basel
Switzerland
Tel.: +41 61 683 77 34

Sensors Editorial Office
E-mail: sensors@mdpi.com
www.mdpi.com/journal/sensors

www.ingramcontent.com/pod-product-compliance
Lightning Source LLC
Chambersburg PA
CBHW061559210926
51850CB00009B/2153
9783725816750